Introduction to
Computer-Aided
Manufacturing
in Electronics

INTRODUCTION TO COMPUTER-AIDED MANUFACTURING IN ELECTRONICS

DOUGLAS A. CASSELL

Control Logic, Incorporated
Natick, Massachusetts

WILEY–INTERSCIENCE, a Division of John Wiley & Sons, Inc.
New York • London • Sydney • Toronto

TO PAT

Preface

This text had its origin in preparations for a course in Computer-Aided Manufacturing (CAM) which I taught at the 1971 National Electronic Packaging and Production Conference in Anaheim, California. The objective of this course was to provide an introduction to the use of computers in manufacturing electronic equipment. The course was directed toward practicing managers and engineers in electronic production.

When I began to collect materials for the course, I found that very little technical material was available in the literature. Virtually all the written material on the subject was in trade journals and consisted largely of chatty little success stories and vague introductory articles. Various conference proceedings contained information of a firmer technical nature, but lacked in applicability to the problems of the new or general user of computer aids to electronic manufacturing. The success stories served only to record the fact that someone had found a computer to be of some value to his manufacturing operation. The introductory articles were generally limited to topics in basic numerical control and concentrated on metalworking and attempts to define the latest acronyms.

All the real technical material was either in the heads of engineers designing CAM systems or scattered throughout a wide variety of engineering textbooks and computer system manuals. The potential user of a CAM system had no choice but to go out and collect (or reinvent) the technical material required for his project.

Beyond the technical matters involved in CAM systems, I have found that many potential users of these systems have no concept of the rewards (and problems) that come along with the use of these systems. Nor do they have any idea of how to procure such a system, select vendors, design the initial system, and so on.

It seemed, therefore, that some need existed for a compendium of basic technical and managerial guidance material on the subject of computer-aided manufacturing in electronics. I hope that this textbook will help to fill this need.

I would like to thank Richard F. Brown, President of Computer Guild, Inc., and Bruce E. Peck, President of Control Logic, Inc., for providing much valuable guidance and encouragement and a fine working atmosphere during this project. I would like to thank my colleagues at Control Logic, especially Warren Mayhew, Paul Gilbreath, Joseph Dzigas and Barry Kover, for many interesting discussions and criticisms. I would also like to thank Mrs. Carolyn Ferguson for her fine secretarial work on this project and for her excellent command of English grammar. Finally, I would like to thank my wife for her patience, encouragement and quietness through many long nights at the typewriter.

DOUGLAS A. CASSELL

Sherborn, Massachusetts
December, 1971

Contents

Chapter One   Introduction   1

Chapter Two   The Electronics Manufacturing Process   4

Some History and Projections, 4
The Phases of Electronics Manufacturing, 6
   Product design phase, 6
   Development of a manufacturing data base, 8
   Fabrication and assembly, 9
   Testing, 10
   Stocking and distribution, 11
Integrating Factors and Control Paths, 12
Distinguishing Characteristics of Electronics Manufacturing, 12

Chapter Three   Tools for Electronics Manufacturing   14

Product Design, 15
   Traditional tools, 15
   Computer-aided design systems, 15
   Simulation, 18
Development of a Manufacturing Data Base, 20
   Traditional tools, 20
   Automatic drafting machines and digitizers, 21
   Parts programming, 24
Fabrication and Assembly, 26
   Traditional methods, 26
   Numerically controlled tools, 27
Testing, 35
Stocking and Distribution, 37

Chapter Four   The Modern Digital Computer   39

Classical Structure and Principal Types, 40
Applications, 48
  Stand-alone systems, 49
  Process control systems, 51
  Tracking systems, 55
  Time-sharing systems, 57
Computer System Architecture, 63
  Classical architecture, 64
  Microprogrammed computers, 73
  Decentralized structures, 76
  Input/output structures, 77
    Programmed input/output, 78
    Interrupts, 81
    Direct memory access, 86
    Input/output processors and channel controllers, 88
    Peripheral devices, 89
  Advanced architectures, 94
    Memory organizations, 94
    Multi- and parallel-processor systems, 97
Computer Languages and Software Systems, 98
  Machine language, 100
  Assembly language, 102
  Procedure-oriented languages, 106
  Software systems, 110
General Capabilities and Limitations—Summary, 112

Chapter Five   The Computerization of Manufacturing Tools   114

Technical Motivations and Objectives in Computerization, 115
General Design Philosophies, 122
  Analog-digital mix, 124
  Computer responsibility, 130
  Data base handling, 144
  Packaging and general design, 151
  Human factors, 154
  Computer requirements, 157
  Program design, 163
Design Example, 171
Summary, 180

Chapter Six   Integration of the Manufacturing Process   182

Management Objectives and Responsibilities, 183
Integration into the Present Manufacturing System, 188
Integration into the Present Corporate Structure, 196
Enhancement of Productivity, 201
Enhancement of Effective Control, 206
Acceptance of the System, 218

Chapter Seven   Getting Started in Computer-Aided Manufacturing   220

Planning, 220
Costs, 228
Procurement, 231

Chapter Eight   Trends   234

The Economic Sphere, 235
The Technological Sphere, 239
The Organizational Sphere, 242

Bibliography   244

Index   245

Introduction to Computer-Aided Manufacturing in Electronics

Chapter One

Introduction

The subject of computer-aided manufacturing (CAM) covers the application of computers to every phase of the manufacturing operation. Briefly, CAM is the discipline of applying computers to the manufacturing process: for the control of machinery, for handling the data that describe the manufactured products, and for the control and handling of material flowing through the manufacturing process. The discipline of computer-aided design (CAD) is both a part of CAM and complementary to it. Computer-aided design is the discipline of applying computers to the generation of data that describe the products being manufactured. The output of the CAD process represents a major part of the input to the CAM process, and the characteristics and outputs of the CAM process influence the operation of the CAD process feeding it. The scope of both disciplines covers every manufacturing volume, from the fabrication of prototypes to assembly line production of many copies of the product.

Many corporations, under the pressure of rising manufacturing costs and increasing competition, have sought means of reducing operating costs and increasing responsiveness to the requirements of their customers. The application of computers, in various ways, to their manufacturing processes has often been found to be a way of satisfying these objectives. Computer-aided manufacturing is a modern and dynamic application of computers. Computer aids and direct computer control are employed in virtually every phase of electronics manufacturing to enhance productivity and flexibility and to reduce costs and turn-around time.

The first use of computers in manufacturing automation did not lower operating costs but rather improved other factors, such as product quality and uniformity, employee working conditions and safety, and the general problem of monitoring manufacturing operations and providing tools for

1

more effective control. Recent technical trends in the field of data processing and computer control have introduced new factors and resulted in dramatic changes in some of the characteristics of these applications, particularly in regard to economic considerations. A principal contributor to these changes has been the small, general-purpose, digital computer (christened the "minicomputer," a term now in general use throughout the industry).

The history of computerized manufacturing processes is littered, here and there, with cases of marginal success and both minor and abject failure. In many cases, the reasons for failure are grounded in a lack of understanding of the basic principles involved in the problem, exaggerated expectations based on the glowing promises of the brave new world of automation, and the improper evaluation of true costs versus compensating benefits. Many innovative manufacturing managers have established automated systems that failed because these managers, in their otherwise commendable attempts to put things in perspective, viewed their problems from a distance that obscured important features of their problems and the solutions to these problems. This is another variation on the old "for want of a nail" theme. The details are often as important to a successful solution as the major aspects of the problem. This is particularly true of computer applications. Computers are appallingly literal and precise beasts, and it is not always easy to live with them; but they do provide genuinely useful services which can generally result in improved operations when they are properly applied. When these improvements promise to reduce operating and manufacturing costs, it behooves the wise manager to invest in their potential application.

The first investment is in learning about the capabilities of these machines and the ways in which they might possibly be applied to manufacturing operations. At the present time, and in the foreseeable future, the situation is that CAM systems are not available as "off-the-shelf" products. They cannot be obtained through a simple salesman/catalog purchasing route but must be fabricated in forms especially tailored to the solution of a particular user's problem. The best people for the job of specifying the characteristics and designs of these systems are their users, and their users are electronics manufacturing personnel. It therefore behooves these people to have a continuing interest in the capabilities of computers, the latest technological developments in the fields of computer system design, and the economics of their applications.

The objective of this textbook is to provide the reader with technical and background information and guidelines for the application of CAM in his own organization. We attempt to satisfy this objective by introducing the reader to basic principles of the design, application, and economics of computer systems in electronics manufacturing so that he will fully understand the factors involved in the solution of manufacturing problems through computerization, the tools available for these applications, and the capabilities of these tools.

Chapter 2 deals with the electronics manufacturing process from a generalized point of view and Chapter 3 discusses the tools that are presently available for use in the manufacturing process, some manual and some automated through the use of special hardware (specifically, numerical controls). After this groundwork is laid, Chapter 4 is devoted to a treatment of the characteristics of the modern digital computer. Here, the design and capabilities of computer systems are discussed and illustrated with descriptions of some typical applications. The computer programs (software) and the input and output of data are vital factors in the success of a CAM system. These are treated so that the reader will be acquainted with the basic terminology of these fields. Chapter 5 discusses technical considerations pertinent to the direct control of manufacturing machinery with digital computers. The final chapters are devoted to the managerial and economic aspects of developing and operating a CAM system. The principal concerns here are the integration of an automated system into existing operations and the corporation as a unified whole, the enhancement of productivity and control, and the problems and costs of initial system development and procurement.

Chapter Two

The Electronics Manufacturing Process

The success of a computer-aided manufacturing system rests on the premise that the potential user of such a system "understands" the system that he is using. The computer is a tool and its proper use is predicated on the understanding of the work that it is to perform.

Therefore, to lay the groundwork for the detailed presentation of the aspects of computer-aided manufacturing that is to follow, we first discuss the electronics manufacturing process from a somewhat generalized viewpoint. We describe the various phases of the electronics manufacturing process, the characteristics of these phases, and some of the ways in which they relate to each other. We also cover some of the principal difficulties with present manufacturing methods and the ways in which these difficulties are amplified or diminished by the characteristics of the products being manufactured.

SOME HISTORY AND PROJECTIONS

By virtually any measure that we may apply, the greatest thrust of automation has been in the heavier industries: steelmaking, automobile manufacture, petroleum, and chemicals. In these industries, the motivation to automate has been exceptionally strong. Rising wages and production costs, the increased value of a single manufactured item, the volume and quality demands of an increasingly affluent and sophisticated consumer society, and the greater technical complexity of each product have all created tremendous pressures on the heavier industries. Automation has been sought as a release valve for these pressures.

Heavy industry has a much longer history than the electronics industry. It is represented largely by rather conservative firms, and it has taken many years for a significant amount of automation to find its way into the heavy industries, in spite of the strong pressures for its adoption. Contrastingly, the electronics industry is much younger and has been undergoing a tremendous rate of growth. This growth was built in large part on factors and impulses that did not originate within the industry itself. Among these factors are the large expenditures for national defense, which are traceable to social and political factors arising out of World War II, and the space program, whose impetus is not unrelated to the same factors that generate pressures for defense spending. The modern electronics industry as we know it is only 20 to 25 years old, and a great deal of its growth has been governed by technological innovations that permitted the application of its products in national defense and aerospace exploration. Principal among these innovations are the transistor, integration techniques, and the modern digital computer. As a result, the electronics industry is volatile, lacking a large body of established tradition and being highly captive to markets over which it exerts little direct control.

Futurists are predicting changes in the application of electronic technology, in particular increased use of electronics in the fields of urban development and control, solution of social ills, and satisfaction of general consumer demands. A recent Electronic Industries Association report* shows that, in terms of dollar volumes, these predictions are beginning to manifest themselves. The growth of the electronics industry has declined to the slowest rate in its recent history and, for the first time in many years, sales for defense applications have actually fallen. Sales for industrial applications continue to expand, however, and offshoots from these applications in social areas are expected to contribute more heavily to this growth in the future.

Historically, the electronics industry has not been its own best customer. This is especially true of the application of its products to the improvement of its manufacturing capabilities. For example, the application of computers is vastly greater in the financial and business communities and among heavy industry. We build them, but we do not find as much use for them as do our customers.

A number of relatively new factors are expected to change this. For example, there has been an increase in the number of small firms engaged in the manufacture of low-volume, custom-designed electronic components and systems. Many of these companies are operating on very low capital bases and profit margins. They have more of the characteristics of belonging to a service industry than to a manufacturing industry. Such companies

* *EIA Yearbook—1969*, Electronic Industries Association, Washington, D.C.

have extremely strong motivations to look for ways to cut production costs and lead times. One of the means that they are investigating and beginning to use is computers.

Another factor is the minicomputer. The low cost of these computers promotes their utilization and will continue to do so. Expected future reductions in the costs of minicomputer peripheral equipment will accelerate their utilization. The future should bring about a situation in which many small electronics firms will employ, as end users, several in-house computer systems within a single facility.

THE PHASES OF ELECTRONICS MANUFACTURING

Electronics manufacturing has five major phases, all of which are more or less common to other forms of manufacturing. They are the following:

1. Product design
2. Development of a manufacturing data base
3. Fabrication and assembly
4. Testing
5. Stocking and distribution

Each of these phases will be treated in some detail, but the greatest interest is in phases 2, 3, and 4, since it is there that the digital computer finds its newest and most unexploited applications.

Figure 2.1 illustrates the manufacturing process and some of the control paths that exist between the various phases of the process. It is the feedback and feedforward control paths that provide the integrating factors in the manufacturing process and their proper consideration is essential to the development of an efficient computer-aided manufacturing system.

Product Design Phase

The product design phase begins with the basic system or device concept. At the start of this phase, we know that we wish to manufacture a system or device which performs some known function or functions. We have a rough idea of what the characteristics of this device or system must be, and we have the engineering and scientific knowledge to create a proper and complete "paper design." That is, we have solved any research and development problems that may be associated with the product.

As we proceed, we refine our initial design to obtain a basic design with partitioning into subsystems or components. We decide from what classes of components or devices we will fabricate the device. For example,

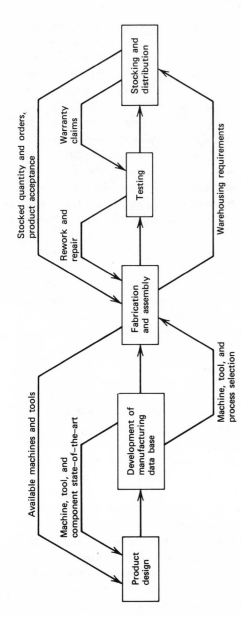

Figure 2.1. The manufacturing process.

7

do we want to use integrated circuits or discrete components? Once the structure of the system or device is laid out, we begin our basic "numeric" design. We select the values of the components; decide the speed of the response times of the circuits; and determine the number of circuits of a particular type or class required. The "numeric" design represents an almost complete and final design of the product.

Finally, we test the design as much as we can. The extent of design testing is governed by several economic and technical questions. For example, is there some area of the design which is especially critical? Is a design change in this area difficult to implement after the prototype has been built? Do we have several different designs from which to choose and are the relative advantages of each unclear? How much will the design test cost compared to the cost of going ahead and perhaps havng to change the design later?

There are several means of design testing. The fabrication of "breadboard" or "boilerplate" versions of the design is one traditional way. A more recent technique is simulation using modern digital computers. A computerized "model" of the design is programmed and exercised on the computer in a number of simulated operational situations. The design is then evaluated according to the results of these simulated situations.

Development of a Manufacturing Data Base

The next phase is the development of a manufacturing data base. During this phase, the design is fully specified in final numeric and graphic detail, and it is therefore an extension of the numeric work done during the product design phase. Layouts, schematics, and scaled drawings of the component parts of the system or device are produced. Parts are assigned numbers, wiring lists are generated, and tapes are produced for numerically controlled machine tools. Artwork masters and masks are produced for the etching of printed circuit boards and the manufacture of semiconductor components by diffusion and deposition processes.

The development of the manufacturing data base is heavily influenced by the manufacturing techniques to be employed. The requirements for numerical detail and accuracy are linked to the capabilities of the machine tools and processes to be used in the manufacture of the device. This is a two-way link. Some aspects of the design can lead to requirements for new manufacturing tools. Conversely, the presence of certain tools already in the plant can influence the requirements of the data base.

A vast amount of labor is invested in the development of a manufacturing data base. To a great degree, the end product of this labor is represented on paper, photographic film, and punched or magnetic computer tape.

Drafting departments must produce drawings showing the part in various stages of assembly and manufacture. Dimensions from these drawings are used to produce jigs and fixtures for machining and drilling operations, and control tapes for numerically controlled machine tools. In printed circuit board and semiconductor manufacture, these accurate drawings are photographed and reduced for use in photoresist and masking processes.

The paper and film produced during this phase of the electronics manufacturing process have value lasting beyond the actual manufacture of the device or system, since they serve as product documentation. This documentation is used during manufacturing and while the device is serving in its final application. Throughout the manufacture of the device or system, this documentation serves as the basis for product testing, quality control, and quality assurance. It is utilized during the diagnosis of malfunctions in the field and in the specification of replacement parts. As systems, of which the device is a component part, proceed through various generations and phases of use and upgrading, this material serves as the basis for modifications, the specification of engineering changes, and the specification of characteristics for replacement parts in the upgraded system.

For the most part, the manufacturing data base is generated by hand through the direct labor of engineers, draftsmen, and technicians working under the supervision of the original designers of the component, device, or system. This represents a significant portion of the total labor expended before a design is finally committed to actual full-scale production. For some complex devices and systems, the development of the manufacturing data base can consume as much as 50 percent of the preproduction labor and dollars.

Among the phases of the electronics manufacturing process, the development of the manufacturing data base is exceptionally amenable to computerization. As we will see, inexpensive minicomputer systems prove to be quite useful in reducing the costs associated with this phase.

Fabrication and Assembly

Actual fabrication and assembly is the third phase of the manufacturing process and comes most immediately to mind when we think of manufacturing: machines grinding out parts by the hundreds and thousands.

Looking at the creation of a complete electronic system from the ground up, we begin with the manufacture of the individual components. These components are destined for mounting on circuit boards that must have interconnections etched and plated or wired. Back panels and frames are wired to form the interconnections between the different circuit boards. Finally, circuit cards are mounted in back panel frames and assembled into full racks of equipment that are interconnected by cabling to form the complete system.

There are many detailed steps in the fabrication and assembly phase, and they vary widely, depending on the device or system being manufactured. In many instances a large number of the basic components and subassemblies can be purchased from outside sources and therefore appear in the process as a procurement step. Masking, deposition, and diffusion processes in the manufacture of discrete and integrated circuit components may involve large numbers of steps. The assembly of certain devices and systems can be incredibly complex. Individual circuit boards may have as many as five hundred to one thousand individual interconnections and hundreds of components can be mounted on these boards. Back panel wiring to interconnect separate logical and functional elements may consist of ten to twenty thousand wires strung within a single rack of equipment. Interconnections between racks of equipment can involve distances ranging from a few inches of light wire to several miles of underground pressurized cable. Physical mounting may range from a single small cabinet to several hundred racks.

It is very difficult to put the complexity and diversity of the manufacturing process into a few words. It is true that there are many products that come off a single linear, branchless, production line consisting of a few steps. However, at the other end of the spectrum we find complex manufacturing processes that depend on widely distributed supply systems for input parts. Parts may also come off sections of such a production process and be stored for several months before they are injected back into the process. Some parts may be diverted at a certain point in the process. For example, components that fail to meet military specification can be diverted to a commercial or industrial application.

The application of numerical control (NC) is primarily in the fabrication and assembly phase. Numerical control has achieved a wide penetration of heavy industry and is already widely used in electronics manufacturing as well. The computerization of fabrication and assembly presupposes the use of numerically controlled machine tools and, as we will see, extends and amplifies the benefits of using numerical control.

Testing

Throughout the fabrication and assembly phase there are stages at which testing should be performed, although we generally think of testing as a terminal activity at the end of the fabrication and assembly of the product. This testing may range from a full examination of the proper operation of every subunit (down to the smallest component) at each stage of manufacture to the spot checking of a statistically selected sample taken from the end point of the manufacturing process.

The amount of testing and the points at which it should be performed are, of course, highly dependent on the product itself. Relatively simple devices manufactured in large quantities are generally tested only at the end of the manufacturing process and then only in randomly selected small lots. As the product becomes more complex, the testing question becomes more complex. For example, contrary to our intuition, as a device (e.g., an electronic instrument) nears the end of its manufacturing cycle, it may actually require fewer tests to assure its proper operation than would be required to assure the proper independent operation of each component in the instrument. However, if the instrument fails the test, the manufacturer is faced with the cost of troubleshooting and repair. Depending on the point at which the tests are made, the cost of isolating a faulty component can exceed the cost of pretesting the components. These are considerations that should be carefully examined and weighed when setting up a manufacturing system. They have impact beyond the cost of manufacture as well, since they influence warranty expense and the general reputation of the product for quality.

Many of the general characteristics and requirements of testing make computerization an attractive alternative to the manual systems that are widely used at this time. Testing generally produces large quantities of data that must be analyzed in various ways to maintain an effective quality control program. The analysis of test data is a job to which a computer is admirably suited. Furthermore, the testing of complex devices frequently requires that they be exercised under a large number of situations or that a large number of individual tests be performed. A computer is often useful in controlling these situations and recording the resulting data.

Stocking and Distribution

At the far end of the manufacturing process, we find a logistical system for the stocking and distribution of the product. This system functions as a buffer between the manufacturing process and the "outside world" and distributes the product to the "outside world" when it is called for. The operation of this logistical system should exert a strong controlling influence on the manufacturing process itself, largely in terms of the volume that the process is allowed to produce. For example, as the stocked quantity of a product fluctuates because of changes in the demand for the product, the volume produced by the manufacturing process should also fluctuate, perhaps turning its capacity to the production of items for which there is a lower stockpile and an increasing demand. In considering the computerization of a manufacturing system, we also consider the influence that the market place exerts on the process and how the system can best sense this influence and respond to it.

INTEGRATING FACTORS AND CONTROL PATHS

No single phase of the electronics manufacturing process stands by itself, unaffected by the other phases. Feedback and feedforward control and influence paths are laced through the whole structures of the process, and their consideration is essential to an understanding of the process. The influence of these feedbacks and feedforwards is illustrated by the following rules—rather obvious rules which are occasionally violated in varying degrees:

1. Do not design a product that cannot be manufactured and tested efficiently and adequately.
2. Do not specify component values and performance characteristics that are not attainable.
3. Do not wait until the device or system is completely assembled or manufactured before testing.
4. Do not make more than you can stock, distribute, and sell.

As we examine the computerization of the manufacturing process, we consider these control and influence paths further.

DISTINGUISHING CHARACTERISTICS OF
ELECTRONICS MANUFACTURING

Electronics manufacturing has several characteristics that distinguish it from more general forms of manufacturing and heavy industry. These characteristics are, in many cases, strong supporting points for the computerization of the manufacturing process.

Electronic systems, in most cases, contain more separate parts than any other form of manufactured product. These parts have a much smaller average size than most other manufactured products and they must be assembled to more exacting tolerances. Their value is also higher than that of the average product component. The principal work of fabrication and manufacturing in electronics is assembly of these parts—there is relatively little in the way of metal removal, milling, and cutting.

The characteristics of the market to which electronics manufacturers sell is also sufficiently different from normal manufacturing markets to exert strong effects on the manufacturing techniques employed. This market is extremely volatile, and its requirements are continually changing. One of the primary reasons for this volatility is the large amount of research devoted to the development of new technologies and techniques. Each new discovery resulting from this research generates new pressures on manufacturers to take advantage of the technology or to provide devices and systems that utilize this new technology. Innovative consumers of electronic products are continually discovering or inventing new applications for electronic products and these applications frequently have characteristics that generate additional pressures on the manufacturer to alter his product in some way which enhances its applicability to the requirements of these consumers. For example, the development and application of the minicomputer and its remarkably low costs have stimulated interest in minicomputers in a large number of market areas which could not have considered their application in the past. Interest grew and, in its turn, generated pressures for even lower costs, additional capabilities in these computers, and lower costs for peripheral equipment.

Finally, the documentation requirements placed on the electronics manufacturer are much more severe than those placed on most other manufacturers. The documentation that generally accompanies an electronic system of average complexity can run to several large volumes and include drawings of printed circuits, wiring diagrams, wire lists in various forms of organization, spares lists, test data, and operation and maintenance manuals that provide correlated analyses of various fault conditions and their causes and cures.

Chapter Three

Tools for Electronics Manufacturing

In this chapter, we discuss some of the tools that are employed in the manufacture of electronic products. Many of these tools already demonstrate a significant amount of. automation, generally because they are controlled by some form of special electronic or electromechanical device. In a few instances, we encounter computerized devices and procedures.

This discussion of electronics manufacturing tools is organized along the phase lines that are described in Chapter 2. Within each phase, we examine the technical operation of particular tools, economic aspects, what kinds of companies use the tools, and the development history of the tools. The manufacturing phases and the principal classes of tools associated with each are listed below:

Product Design
> Engineering labor
> Computer-aided design (CAD) systems
> Simulation

Development of a Manufacturing Data Base
> Digitizers
> Drafting machines
> Artwork generators
> Parts programming languages

Fabrication and Assembly
> Numerically controlled tools
> Printed circuit board drills
> Component insertion machines
> Wire termination machines

Testing
  Component testors
  Printed circuit board testers
  Interconnection testers
Stocking and Distribution
  Inventory control systems
  Management information systems

PRODUCT DESIGN

Traditional Tools

By far the greatest amount of design is done using the traditional tools: direct engineering labor with pencil, paper, slide rule, desk calculator, and manufacturer's catalogs. Even in the most advanced corporations doing electronic design, these tools still find widespread employment. This will be the case for many years to come.

The greatest portion of design costs using the traditional tools is the direct cost of engineering labor. This cost is presently rising at a rate that is bounded below by the national rise in personal income. Engineering labor is expensive and it will probably become more expensive in the near future. New technologies and the "information explosion" tend to amplify the cost of engineering labor; the manufacturer must hire either more specialists or more expensive generalists.

The most unpredictable protion of design labor is general "think" time—the time devoted to invention and development of the concepts on which the product is based. One large portion of design labor is the time spent on engineering calculations: calculating circuit performance, adjusting parameters, developing Boolean equations, Karnaugh map analysis, and so on. The first entry of the computer into the product design area was as a tool for engineering calculations. This leads us to our next category of tools: computer-aided design systems.

Computer-Aided Design Systems

Although their use is concentrated in the larger corporations (particularly computer, semiconductor, and aerospace companies), computer-aided design (CAD) systems are already in large-scale use for electronics work.

The precise definition of a CAD system has been, and still is, the subject of a small controversy. Until recently, some widely accepted definitions of the term included activities in direct automation of manufacturing. The latest definition restricts CAD systems to those directly applied to the first phase of electronics manufacturing, as we define it in Chapter 2. For our purposes, the following definition will apply:

A computer-aided design system is any collection of hardware, software, and procedures that involves the utilization of a computer in the activities of component, device, or system design up to, but not including, the production of representations of any portion of the design that participate directly in the actual manufacturing process.

According to this definition, we see that programs for the solution of simultaneous linear equations (used to calculate voltages and currents in passive networks) qualify as CAD tools. However, the computer-aided cutting of a mask for integrated circuit etching does not qualify as a computer-aided design tool. Processes involving the use of a computer that do not fall within the limits of this definition belong to the category of computer-aided manufacturing (CAM) systems—the central subject of this text.

There are strong arguments for treating both kinds of systems as a unified whole, particularly when considered from the viewpoint of the manufacturer interested in the integration of the complete manufacturing process, from concept to delivery. The present division of responsibility is due in no small measure to the fact that design and manufacturing techniques have respectively been the bailiwick of two separate groups: the academic and the industrial communities. Computer-aided design originally appeared in the academic community, which was dominated by design-oriented people. Computer-aided manufacturing (and automation, in general) originated within the industrial community, which was dominated by production-oriented people. The emergence of integrated electronic manufacturing companies has brought the two related technologies together, but the present division of responsibility still holds.

There are several kinds of CAD systems, and they are available to the engineer in degrees that vary with the size of the corporation at which he is employed, its farsightedness over the last decade, and its willingness to invest funds in nonproduct research and development. There are exceptions to this, of course, and they are primarily due to the offerings of several commercial computer time-sharing organizations.

The first CAD systems were computer programs that were used to calculate the performance of passive networks. These were programs that solved sets of simultaneous linear equations. For the most part, they were borrowed from mathematicians, and the names of specific programs are lost in the early history of computing. These programs solved well-known engineering problems that required handling large numbers of variables in an accurate and fast way. They incorporated little human engineering, often required the combined talents of a mathematician and a computer programmer, and produced a collection of cryptic numbers for output.

The next generation of these programs extended their capabilities to the analysis of active circuits and simplified the process of preparing input. In the earlier programs, the engineer was required to formulate explicitly all the equations that described the behavior of the circuit under analysis. In the new programs, the engineer could list the elements in the circuit and specify their values and the other elements to which they were connected. The computer programs were "intelligent" enough to figure out the appropriate equations and solve them for the variables involved. By this time, plotting devices were becoming familiar computer peripherals, and it became possible for the engineer to see his results in graphic form.

One of the most prominent of these programs was the Electronic Circuit Analysis Program (ECAP), developed by the International Business Machines Corporation and widely used today in various forms as a component part of advanced CAD systems. Now there are several types of circuit analysis programs, all offered by major computer manufacturers or software distribution organizations such as Computer Software Management and Information Center (COSMIC), an activity of NASA and the University of Georgia.

A number of CAD programs are available for use in other areas of electronic design, particularly the design of filters, analog systems, and semiconductor masks. For example, some programs provide frequency response and phase response plots when given the complex transfer function of a filter or pole-zero data. Others allow the user to describe his analog system in terms of integrators, gains, summing points, and so forth. They then calculate the time-domain and frequency response of the system and print or plot the resulting data. Almost every major semiconductor manufacturer uses CAD systems in laying out their masks. Large CAD systems were, for the most part, developed in-house by their users or contracted to software firms.

We mentioned "time sharing" earlier. We cover the details of time-sharing systems in Chapter 4 and explain their basic principles of operation. However, for now, note that time sharing is a method of using a computer in which the individual users are given access to the computer in a way which makes it seem that they have continuous use.

Commercial time-sharing organizations maintain one or more large central computer systems and sell time on these systems to their subscribers. Subscribers to a time-sharing service generally communicate with the computer using Teletype terminals. Commands and input data to the computer are entered through the Teletype keyboard. The output from calculations performed by the programs is printed on the teleprinter. This communication generally takes place over public telephone lines. Since many users are sharing the same computer, the costs of its operations can be distributed over all the users, resulting in a greatly reduced cost to the individual user.

To attract subscribers, commercial time-sharing services develop systems of software for use by their customers. We frequently find CAD programs, of the kinds described earlier, available on these time-sharing systems. The most popular of these are the circuit analysis programs, such as ECAP.

A great deal of progress has been made in computer-aided design over the last decade, and programs are now available that have application in almost every area of electronic design. However, there are still many problems to be solved. The design activity can be broken into the activities of synthesis and analysis. All the CAD programs that have been mentioned are primarily oriented to the analysis end of the design activity. Programs that actually synthesize a design are rare and are generally still in an experimental stage.

Another problem is making CAD systems available to a larger number of engineers and making them easier to use. The solution of this problem is related to the development of computer terminals, improvement of the capabilities of time-sharing computer systems, and development of better languages for communicating problems to the computer.

Simulation

Assuming that the economics of the situation justify its use, simulation is one of the best ways to test a design before committing it to hardware. The general method of testing a design by simulation consists of representing the design by a programmed model. This model is then exercised on a computer with input data that represent operational situations of interest.

In the case of circuits, the relevant equations constitute this model and the analysis of the circuit using the CAD systems just described is equivalent to simulating the circuit. The theoretical borderline between CAD systems and simulation systems is rather hazy. At the point when a design-aid system is more properly considered a true simulation system, we begin to deal with large, complex systems or systems that must operate in a complex and occasionally indeterminate environment. One example of such a problem is the testing of the design of a telephone switching network. Requests for system service and the loads they present to the system are random in nature and their analytical treatment is frequently rather difficult.

While system models can be programmed from the ground up, several new computer languages have been developed. They allow the designer to represent his system in terms of functional blocks or subsystems, whose characteristics have already been represented in parameterized form in the computer programs that translate these languages. These languages also permit the designer to vary the exercise situations over many cases automatically and to organize the data that result from the simulation in the way most convenient for analysis or presentation. Most of these simulation systems also allow the designer to call for pseudorandom events so that he can test the statistical behavior of his design. Some examples of these simulation systems are the General Purpose System Simulator (GPSS) and the Continuous System Modeling Program (CSMP).

To a great degree, simulation systems have about the same distribution of use as the CAD programs discussed earlier. That is, they are principally employed by the larger electronic companies. Many of these programs were originally developed by universities and research firms to study economic and military systems.

Of more interest to the electronics manufacturing industry are programs that simulate digital hardware. These programs utilize system models built from either Boolean equations or block-diagram models in which the fundamental units are gates, inverters, flip-flops, and so on. The designer describes each component of his device or system in terms of function, response time, fan-in, fan-out, thresholding, reshaping, and others. He then specifies the way that these components are interconnected. After specifying a number af excitation waveforms or inputs, the designer can instruct the program to exercise the design. The outputs of the program show resulting waveforms at specified connectors, logical states of the system at various points in time, and so forth. One of the most important uses of these simulation programs is in the examination of overload conditions, the degradation of output and synchronizing waveforms, and the detection of logic errors.

Some digital system simulation programs are available from commercial time-sharing services. However, almost all these programs suffer from difficulties in their operational use. First, the amount of input required to describe adequately a large digital system is rather voluminous. Second, the calculations involved are frequently extensive and the amount of computer output is somewhat large. This is especially true of programs that report on waveform degradation and include an analysis of system timing. This is a serious technical difficulty and is due to the fact that a computer is a sequential processor, as we shall see in the next chapter. The computer can handle only one situation at a time, which makes it particularly difficult to analyze digital systems in which many events are occurring simultaneously.

DEVELOPMENT OF A MANUFACTURING DATA BASE

Traditional Tools

As we noted in Chapter 2, a great deal of the work in developing a manufacturing data base is being done with manual methods. The principal output of this activity is most often a graphic or numeric representation of the product design. Graphic representations of the design are wrought using the arcane arts of the skilled draftsman. Numeric representations are generally conveyed by dimensioned drawings and wiring lists.

As the device becomes smaller in physical size, the requirements for graphic and numeric accuracy become more severe. The general tedium of graphic representation grows with the increasing complexity of the device. In the electronics industry, the best example of this effect is the making of masks for the manufacture of integrated circuits.

In this phase, as in the product design phase, the greatest cost allocation using the traditional tools is the direct cost of labor, generally drafting and technician labor. As with the design engineer, the manufacturer is finding that the cost of draftsmen and technicians is increasing, while their availability is going down. Many technical schools which originally guided their students into drafting are now tempting them with careers in computer programming.

After laying out masks for integrated circuits and printed circuit boards, the second greatest cost allocations for this phase are found in the generation of system and subsystem logic schematics and wiring lists. The presentation of logic schematics is not a particularly demanding discipline when compared to mask cutting, since a dimensionally precise drawing is not required. For a complex system, it can be a rather time-consuming process, because a good deal of detail and lettering is required. This is alleviated somewhat by the transparent stick-on drawings of logical elements that can be obtained from manufacturers.

One of the major problems is the maintenance of "as-built" drawings and schematics of the system. Changes in the design frequently result during the fabrication, assembly, and testing stages. These must then be communicated to the drafting department and numerous errors are often the result.

The preparation and maintenance of wiring lists also consume a great deal of time. Manual methods that are used call for the assembling of wiring lists from information represented on the logic schematic. For relatively simple systems, the list is simply printed on several sheets of paper and given to a technician or wireman for manual wiring. For more complex systems, the wiring list is frequently punched on Hollerith cards so that simple data processing equipment (called "unit record" equipment in the data processing trade) can sort the lists and print them in easily readable form. Again, the problem of maintaining as-built wire lists is a serious one.

As can be seen in the following pages, a number of automated tools are now in widespread use that provide certain advantages over the traditional tools.

Automatic Drafting Machines and Digitizers

The advent of integrated circuit (IC) technology brought with it strong motivations for the use of automated techniques. The technology of discrete component printed circuits required artwork of relatively low-precision requirements and complexity. Masks for the manufacture of integrated circuits had to be much more precise, and the increased complexity of the single IC device made the drafting problem more difficult. Now, four to eight logical functions are represented on an ordinary dual-in-line package (DIP), the printed circuit boards on which they are mounted have gone to multiple layers, and the dimensions of a single board have increased to the point where the central processing unit of a small computer can now be packaged in a single board without straining the state-of-the-art. In fact, this is now common practice among manufacturers of minicomputers.

It is now the case where most manufacturers of integrated circuit components use automated drafting machines for the production of artwork. Many manufacturers of printed circuit boards also employ these machines, but not to the extent that they are employed in the integrated circuit industry.

The drawings produced by these machines are graphic representations of the device; they represent it to varying degrees of dimensional and symbolic accuracy which depend on the device and the details of the manufacturing process. For example, logic diagrams and schematics are symbolic representations of the device and do not participate directly in the manufacturing process. They are essentially documentation according to which the device will be fabricated. They must be logically correct, but they do not have any dimensional requirements placed on them. On the other hand, drawings from which integrated circuit and printed circuit board masks will be made are accurate pictures of the device in various stages of the manufacture. The dimensions of the various parts of the device and the tolerances obtained in its manufacture are directly related to the dimensions of the drawings. Inaccuracies are represented in the finished product. The automated devices available for the production of drawings are generally tailored to one of these two applications.

High-accuracy drafting machines work from a numeric representation of the drawing, moving a drawing head over some receptive material such as paper, film, or mylar. We cover the methods by which this representation actually drives the machine when we discuss numerically controlled tools later in this chapter. The basic operating principles in moving the drawing head are virtually the same. The drawing head is generally either a pen (for drawing schematics), cutting or scribing tools, or a light head (for making circuit board or IC masks). The drawing is made directly on paper or by exposing high-resolution film or photographic plates. Another technique is cutting of mylar sheets coated with filtering material. Mechanical motion cannot easily be controlled to the resolutions required by the inner dimensions of complex integrated circuits. Thus finished IC mask drawings must be photoreduced to the actual size of the circuit.

Automatic drafting machines are available with drawing areas up to 48 by 60 inches in which drawings can be made with resolutions of 0.001 inch. Photoreduction of drawings made on these machines yields integrated circuit chips with dimensions roughly 120 to 240 mils square. This is the approximate dynamic range of the drafting process. There are, of course, considerations of photographic accuracy and resolution, but they do not directly influence the aspects of automating these processes. Geometrical considerations, on the other hand, are of great importance, particularly the relative proportions of shaded to unshaded areas and single lines of widths available on the drawing head. Shaded areas are filled in by repeated strokes of the drawing head. With lines, apertures or pens of particular widths can be selected and the line is completed by a single sweep of the machine. The drawing speeds that can be obtained from these machines are heavily influenced by the geometrical considerations. Thus machines that rapidly shade in rectangular areas in special raster scanning modes are very useful for IC masks.

Machines for high-resolution artwork generally cost about $90,000. For the preparation of dimensioned drawings, schematics, and other documentationlike material, a number of relatively inexpensive devices are in use. These are generally digital plotters, often found as computer peripheral devices. They may be controlled by NC units or computers. The general nature of documentation drawings is that they are collections of symbols (including letters and numerals) connected at various points with straight lines. In preparing instructions for one of these machines, the operator specifies the symbols, the location of each symbol, and the end points and break points of the lines connecting the symbols. The exact method by which he does this is a function of the machine being used. There are three basic means of preparing input. One method is digitizing a rough drawing. We discuss this shortly.

The second method consists of representing the drawing in a computer language designed for this purpose. A computer program then prepares the detailed machine instructions. With the third method, photographic techniques are used. The operator of the machine selects a symbol from a "library" of symbols which is generally held within the machine on a roll of exposed and developed film. He selects a symbol from this film by depressing a button on the machine console over which the symbol is depicted. This action displays the symbol at a location that he has selected and exposes the symbol to a film upon which the final drawing is produced. Enlarged prints of this film are used to produce the final drawing. Line segments are drawn on the film in much the same way. Generally, the operator specifies positions on the drawing by a cursor or pointer, the position of which is encoded either mechanically or electrically.

Digital incremental computer peripheral plotters are less expensive than the drafting machines mentioned above in terms of hardware investment, but they require supporting computer programs that interpret the user's drawing data and produce the plotter pen motions that draw the symbols in the library.

As we noted earlier, many automatic drafting machines are numerically controlled and the control units that operate these machines are similar to those that are used with machine tools. The basic inputs to these are a series of line end points and drawing head commands. The numbers that represent the line end points are generally punched on paper tape, mylar tape, or computer cards, using machines called digitizers. These machines can be used in the preparation of tapes for the operation of most numerically controlled machines. The components of a typical digitizer are shown in Figure 3.1.

The procedure for preparing tapes using a digitizer begins with the engineer or technician drawing up a representation of a printed circuit board (e.g., in rough form on paper on which a precise grid has been drawn). He aligns all the board lands, holes, and interconnection lines with the grid. The time required to produce this rough drawing is, of course, much less than would be required to produce a finished drawing suitable for use as an artwork master. This drawing is mounted on a digitizing table. Each digitizer has a procedure for establishing the zero coordinates at some point on this grid. The operator then uses a reticle which he places over the point to be digitized. The position of this reticle is encoded by the digitizer through a mechanical, electrical, or magnetic linkage. When the operator has the reticle properly aligned, he depresses a button or foot switch that commands the digitizer to punch the encoded coordinates on tape or cards. Digitizers are generally equipped with controls that scale these coordinates to the final drawing size. Special keyboards are used for punching special machine commands. These commands raise or lower the tools, select different line widths, and so on.

Parts Programming

We have just seen how a digitizer can be used to prepare a control tape for an automatic drafting machine. Digitizers are also frequently used

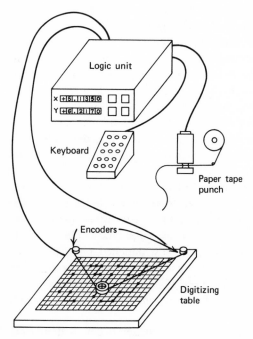

Figure 3.1. A typical digitizer.

to prepare tapes for the control of NC machines used in the manufacture of electronic equipment. The most frequent use of digitizers is in the preparation of programs for automatic component inserting machines and printed circuit board drills.

Of course, the use of a digitizer in NC program preparation requires a prepared drawing either on gridded or accurately dimensioned paper. If the numeric values of the coordinates of drilling or insertion points are known, the control tapes can be punched manually using either a flexowriter or a teletypewriter. In this case, a drawing is not required. All that is required is a list of the points at which holes are to be drilled.

It is frequently the case that the calculation of these coordinates and their translation into positioning commands for an NC tool are not trivial. Some NC machines use incremental positioning commands that instruct the machine to proceed a certain distance from the point at which it is presently located. Others use absolute positioning commands in which a point is specified with respect to a fixed reference point. In the case of the absolute machine, the parts programmer need only enter the coordinates of the drilling points. In the case of the incremental machine, he must subtract the coordinates of the present point from those of the next point in order to determine the positioning command to be punched on the tape. We can see that there are going to be part programming jobs in which the programmer would like to have the assistance of some kind of automatic calculator in preparing the program.

For many years, this function has been served by the ordinary desk calculator. This is still the case in many firms manufacturing electronic equipment. In the metalworking industry, the parts programming job was several orders of magnitude more difficult. It was recognized early that digital computers would have to be employed to help with parts programming. This can be seen clearly from the fact that many problems in contour milling require the generation of thousands of machine commands to describe the infinitesimal machine steps taken while cutting a long curve. The first development in this area was the invention of a computer language called APT (automatically programmed tools). This language was developed in parallel with the first numerically controlled machine tools at the Massachusetts Institute of Technology in the early and mid-1950s. The APT language is now standardized and is maintained by the Illinois Institute of Technology Research Institute (IITRI). Various subsets of this language and special languages for point to point work have evolved over the succeeding years.

Since the great majority of NC work in electronics manufacturing involves point to point work, wherever NC languages are used in the industry they are generally of the simpler point-to-point variety. Because they do not require large amounts of computation time or data, they are available on small computers and time-sharing terminals. One of the best examples of these languages is the QUICKPOINT language, developed by the Digital Equipment Corporation for use on its PDP-8 series of minicomputers.

Until a few years ago, the worst part programming problems that were encountered in the electronics industry were those for printed circuit board drills and component sequencing and insertion machines. Wiring was generally done by hand. If the volume of wiring was extremely large, the wiring work would be subcontracted to companies that had access to the large, fully automatic wiring machines, such as the Gardner-Denver machine. These companies employed computers to set up the NC programs for their machines, but, since only a few companies actually operated these machines, the use of computers for this purpose could not be considered to be a widespread characteristic of the industry. Then the numerically controlled semiautomatic wiring machine came into use. In terms of computational complexity, the programs for these machines were no more difficult than those for printed circuit board drilling and component handling. However, even a system of modest proportions has many more wires than components, and the number of calculations required to prepare a single program increased rather drastically. Furthermore, the order in which wires are to be laid is important. For wrap-type terminations, lower pin levels must be wrapped first. The order in which wires of different lengths are taken will affect the way in which they can be dressed. Thus the wiring lists need to be sorted in various ways. Computer programs for the preparation of these NC programs are now widely used.

FABRICATION AND ASSEMBLY

Traditional Methods

The traditional methods of fabrication and assembly are characterized by manual handling. In some cases, they represent the most cost-effective means of manufacturing. This is particularly true of low-volume specially packaged devices in which there are few components that are of widely varying physical dimensions. Electronic instruments are a good example of this kind of device.

Manual wiring and component insertion are widely used, even for relatively complex devices. One manufacturer of minicomputers, for example, uses a work bench production line for component insertion. Each worker has a cutout cardboard mask that is placed over the printed circuit board. This mask exposes only points at which a particular component is to be placed. Components are selected from a tray and inserted in the board until all the mask cutouts have been filled.

Resistance to the replacement of manual methods has been largely due to the desire to keep capital investments down. However, semiautomatic wiring machines and component insertion machines are beginning to replace manual methods, especially for the assembly of digital systems. Vendors of these machines claim that manual methods are more expensive in the long run and they present very convincing arguments. Since semiautomation requires less capital investment than full automation, their products have proved to be attractive to many manufacturers.

These machines are generally numerically controlled. This leads us to our next topic: numerically controlled tools.

Numerically Controlled Tools

Numerically controlled machine tools are the heart of the computer-aided manufacturing system. We will now investigate the way that they work and some aspects of using them.

Looking closely at the fabrication and assembly phase, we see that jobs can be divided into two principal classes:

1. Situations in which we want to move a tool to a particular point and do work at that point, when we get there
2. Situations in which we want to move the tool along a prescribed path, doing work while the tool is moving

Work falling into the first class is called "point-to-point" work and is the most prevalent kind of numerically controlled work in the electronics industry. Work in the second class in which the prescribed path is a straight line is also point-to-point work and is the kind of work done by a drafting machine. In the metalworking industry, this is generally typified by straight-cut milling. Other metalworking operations generally fall into the second class and, when the prescribed path is curved, we have a class of work that is called "contouring."

Since we are concerned with electronics manufacturing, we concentrate on NC tools intended for point-to-point work. The two most visible parts of such a machine are the tool head and a positioning table. In most cases, the tool head remains stationary while the positioning table holds the work piece and moves under the head. The position of the table is changed through mechanical linkages to numerically controlled actuators. These actuators are generally either direct-current motors or stepping motors. With the former, the velocity of the motor is proportional to the amount of current applied. With stepping motors, the armature rotates through a prescribed angle each time the motor is pulsed. A typical stepping motor configuration is shown in Figure 3.2.

These motors have from four to five leads (depending on the type) on which direct current potentials are applied. Depending on which leads are energized, separate coil groups within the motor are energized. As these groups are energized, they cause the armature to rotate from what was a stable position into a new stable position. The direction of rotation is determined by the changes of the levels at the motor leads. To generalize somewhat, we can say that the problem of driving these motors is one of presenting certain binary switching patterns to the motor leads. The logic used to accomplish this function is some form of shift register or ring counter with two inputs. The first input is a fixed level that sets the direction in which the pattern will switch and, therefore, the direction in which the motor will rotate. The second input is a clock pulse or level change. A pulse or one-to-zero transition on this line triggers the pattern change and this, in turn, causes the motor armature to rotate one position. With a numerically controlled tool using stepping motors, the driving problem at the actuation level consists of direct position control. One pulse or level change presented to the logic causes one motor rotation increment

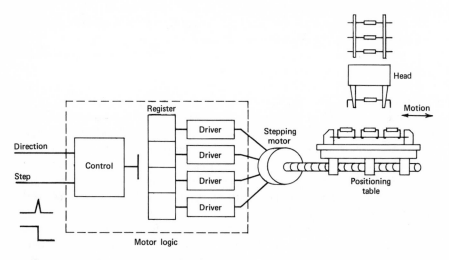

Figure 3.2. Stepping motor driving.

which is translated mechanically into a single linear increment in the position of the table along one axis.

With the direct current motor, usually an industrial-grade servomotor, rotation of the armature starts when current is applied to the motor, and the direction of rotation is determined by the direction of the applied current. The velocity of the motor is proportional to the applied current. With a numerically controlled tool using servomotors, the driving problem consists of direct velocity control.

We can see that the kind of actuator used has an important influence on the control unit itself. Let us look at the problem of moving a positioning table to a certain prescribed position. With either actuator, we must know the current position and the distance from this position to the desired one before we start the motion. If we are using stepping motors, we know

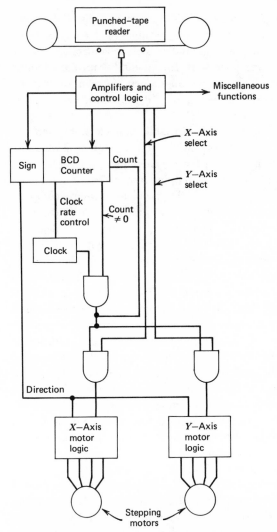

Figure 3.3. Typical stepping motor numerical control, incremental, open loop.

the gear ratios of the mechanical linkages between the motor and the table and we therefore know what linear motion we will get for each rotational increment of the stepping motor. The problem, then, is to subtract the present position from the next position, divide the result by the linear motion that results from a single motor step, and pulse the motor this number of times. Here we have essentially a simple arithmetic operation and a counting operation. Part of the arithmetic operation is done by the parts programmer when he prepares the NC tape. He expresses the command in terms of motor steps. Stepping motor numerical controls come in two basic types: absolute and incremental. Absolute controllers handle the subtraction *and* the counting operation. Incremental controllers handle only the counting operation. The programmer does his own subtraction and expresses the NC command in terms of the difference between the next position and the present position.

Most stepping motor controllers use a variable rate clock to step the motor so that it will not be overdriven when starting up or halted too abruptly, since this can lead to missed steps or overstepping.

Figure 3.3 shows a simplified control for stepping motors.

If we are working with a servomotor, the problem is more complex. If we knew the velocity of the motor very accurately as a function of the applied current and if we were able to apply a precisely measured current, then it might be possible simply to apply the current to the motor for a precisely measured amount of time. Essentially, we would be integrating the commanded velocity of the motor. The value of this integral would tell us when to slow down and stop the motor. However, because numerically controlled work requires very accurate and repeatable positioning and since we cannot count on the motor to respond in a predictable way to the applied current (which cannot be precisely controlled), a different approach must be adopted. Figure 3.4 illustrates this approach. We measure the actual position of the table and compare it to the commanded position. The amplitude of the current applied to the motor is made more or less proportional to this difference. Thus if the table is a great distance from the desired position, a large current is applied and the motor is driven to run at or near its maximum speed. As the table approaches the desired position, the current is gradually reduced until the desired position is reached, at which point the current is shut off and the motor stops. This is a standard first-order servocontrol problem and would be very nice if we could expect a perfectly responsive motor, but, Newton's laws being what they are, the motor and table will continue to move after the current has been shut off and the control system will have to act to compensate for the overshoot. Various modifications to this procedure are used to assure precise control without overshoot or "hunting" (oscillating back and forth past the desired position), such as providing a "dead band" near the zero-error position.

What are the relative advantages of these two kinds of actuators? The servomotor has many characteristics that have led to its preeminence in numerical control, among them are the high torques and speeds of which these motors are capable. Recent improvements in stepping motors that utilize hydraulic amplification have brought them into serious contention

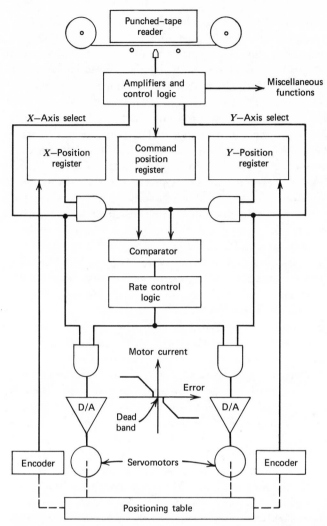

Figure 3.4. Typical servomotor numerical control, absolute, closed loop.

for use in high-torque and high-speed applications, but their use is far from widespread. However, the stepping motor does enjoy some functional advantages over the servomotor. Principally, the control scheme required for stepping motor operation is less complex than that required for servomotor operation. Lower digital logic costs made possible by integrated circuitry have brought the cost of digital controls to a point where they compete with analog controls. We also notice that the control schemes for stepping motor control need not measure the position of the table directly; that is, we do not require closed-loop operation. It is true that we can overload the stepping motor, so that it may fail to take the commanded number of steps, but this is generally an easily avoidable situation. Another advantage is that digital devices do not require the tuning and alignment that analog devices often require.

For both actuators, the commands to the control unit are generally represented on punched tape, either paper or mylar. There are several formats and codes used for these tapes, but the principal codes are either the EIA RS-244 standard or the American Standard Code for Information Interchange (ASCII). The ASCII code was developed as a standard for information processing and communication purposes and is principally associated with computers. It has come into prominence as a result of the increasing use of computers in the preparation of NC programs. The EIA code is much older and is more widely used at the present time.

The two principal tape formats are the Word Address-Absolute and the Tab Sequential-Incremental. There are arguments for and against each kind of format and code, principally centering around questions of programming ease and control unit complexity.

With respect to programming, the Word Address-Absolute format is generally thought to be the easier of the two. If the parts programmer knows the coordinates of the points at which work is to be done, he merely punches them on the tape, preceded by a single character which identifies them; for example, X14301 to specify the x-coordinate. The coordinate itself is expressed in terms of machine resolution; for example, 1 for 0.001 inch. If no change is required in one of the coordinates from one step to the next, it is simply omitted.

With the Tab Sequential-Incremental format, the coordinate is identified by its position within the format, hence the name "Sequential." The first coordinate is generally the x-coordinate. This is followed by the y-coordinate, and if the machine has three axes, then the z-coordinate. With this format, successive fields are separated by the tab character. Tab Sequential formats generally express the table positions in incremental form; namely, the number of steps from the present position to the desired position. Here, ease of programming depends on the data from which the parts programmer is working. For example, if he is working from a drawing of a printed circuit board which is to be drilled and if the hole positions are arranged in a regular array with their separations noted on the drawing, then he can punch these separations on the tape. They are already in the incremental format required. On the other hand, if the hole coordinates are all represented with respect to a fixed point on the board, he will have to do some arithmetic.

As we have said, the widespread utilization of numerically controlled machine tools in electronics manufacturing is a relatively new phenomenon. It resulted largely from needs generated by the growth of the electronics market and the attention of a number of relatively new companies whose principal products are directed to this segment of the electronics manufacturing industry. Almost all major manufacturers of electronic devices and systems requiring a significant degree of assembly now utilize one or more kinds of NC equipment, either in-house or through services provided by either the manufacturers of this equipment or local service companies.

For the most part, these systems were fabricated from lightweight, two-axis positioning tables. Both direct current and stepping motors are used in these systems. To a great degree, the control units were obtained from established manufacturers of numerical controls and retrofitted to the positioning tables. Almost all are low-torque, high-speed, point-to-point machines.

What motivated the electronics industry to use numerically controlled tools? The reasons are nearly the same ones that made NC popular in the metalworking industry. The reason for the lag between heavy use of NC in metalworking and heavy use in electronics manufacturing is probably that the NC units for metalworking were designed for heavy work and were too expensive and slow for the fast, light work of electronics manufacturing.

One of the principal reasons for adopting NC machines is that they work faster than people. Of course, some additional work must be invested in preparing the machine for operation, but it is generally relatively minor. In metalworking, there are some parts that cannot be made *without* NC. This is less true in electronics, where virtually every assembly and wiring operation can be done with manual methods. Here, then, the arguments are generally economic.

A second principal reason is the inherent repeatability of numerically controlled operations. The machine works from an unambiguous program and each part produced is identical to the ones that preceded it and the ones that follow it. The only point at which this will fail to be true is in the event of an error—an electrical or mechanical malfunction.

A third reason for adopting NC is that the process is inherently a low-error rate process. Again, barring electrical or mechanical malfunction of the machine or control unit, the machine will do exactly what it is told. Once a program has been developed that produces the first correct part, it produces all subsequent parts in the same way, with no errors. This claim cannot be made for manual operations.

The two principal reasons have several effects that add to the advantages of NC equipment. For example, fewer errors mean a lower scrap cost and reduced inspection and testing costs. The speed of NC machines means faster turnaround and reduced costs for low-volume runs of a particular product. All these effects create increased profits for the electronic manufacturer.

The chief applications of numerical control in electronics manufacturing are artwork generation (which we have already discussed), printed circuit board drilling, component insertion, and wire termination.

Numerically controlled printed circuit board drills are in use in a variety of configurations, but typical units may have up to four spindles. Boards can be stacked on these machines in groups of four to six boards, one group per spindle. Typical drill stroke times are on the order of a second so that within one stroke the drill can produce from sixteen to twenty-four holes. The long-term output rates of these machines are limited by the stroke time, table positioning time, and the board handling time (the time required to remove one set of drilled boards and replace it with a set of undrilled boards). For typical circuit boards, positioning times vary from half a second to one or two seconds so that once the boards have been mounted, drilling rates of 500 to 800 holes per minute are achieved.

Numerically controlled component insertion machines are also available in a wide variety of configurations. For handling discrete axial-lead components, they generally take the component from a presequenced tape roll. As the positioning table is moved to place the component position under the insertion head, the leads are trimmed and formed. When the table is in position, the component is inserted and the leads are crimped. Component insertion machines are now available for handling DIPs as well as discrete components. There are also a number of machines that come with multiple heads so that boards can be prepared in parallel. Cycle times for each operation are on the order of one second or less, and positioning times are about the same as for printed circuit board drills. However, boards cannot be stacked one on top of another, as they can with drills. Thus, excluding board removal and mounting time, rates of approximately thirty to fifty insertions per minute are achieved. With drilling and component insertion, mechanically programmed machines are feasible, if the production runs are sufficiently long.

Fully automatic wire termination machines have been available and in widespread use for several years. The principal example of these machines is the one manufactured by Gardner-Denver. They cost from $175,000 to $250,000 each and are capable of wiring rates of 500 wires per hour. They are quite complex and the preparation of NC programs to drive the machines frequently requires the use of a computer. They are principally found in large manufacturing facilities which wire large numbers of identical boards.

A relatively recent product on the electronics market is the semiautomatic wire termination machine. This machine acts essentially as a "pointer." The numerical control unit positions the pin to be wired under a wiring head or places it under a reticle. The machine operator then selects a prestripped wire, inserts it in a wiring gun, and actuates the gun manually. The NC control unit then moves the board to place the next pin in position.

There are six or seven manufacturers of these machines and approximately fifteen different models. Most models have a single wiring head and use one operator. There are several that come with tandem heads and require two operators working in parallel. With some of the units developed in-house and some of the less-expensive models, the pin is placed under a reticle and the wire is attached with a hand gun. The board is generally mounted in a near vertical position. Most machines use table motion in two axes, but with some models only the head moves. There is one model in which left-to-right motion is obtained by moving the table, while front-to-back motion is obtained by moving the head. Almost all semiautomatic machines include wire storage bins for holding prestripped wires. Wire selection is controlled from the NC program. The program is set up to display a number that corresponds to a particular bin or to illuminate a light over the bin containing the next wire.

As with the selection of any semiautomatic machine, human engineering considerations are important and influence the wiring rates that can be obtained. Manufacturers of these machines claim rates of between 150 and 450 wires per hour. They represent a significant improvement over manual rates and, since these machines are relatively inexpensive, the wiring costs are generally reduced. Extremely large volumes are not required to justify them economically.

TESTING

The most versatile and widely used test instruments are still the manual ones; the voltmeter, ammeter, oscilloscope, signal generator, and bell and probe used for "ringing" out wired panels. They are the principal troubleshooting devices for almost any electronic system and are widely used by smaller manufacturers of electronic instruments for production line testing. Larger electronic systems are generally checked out with such devices in the later stages of assembly. This is especially true of systems that have low-production volumes, such as some major military hardware.

There are some manufacturers of devices and components who, for reasons real or imaginary, cannot afford to invest in the automated testing devices that are now available. In general, these manufacturers test only a small, randomly selected, sample of their production output. The objective of this testing is not to prevent the distribution of a faulty device so much as to catch breakdowns in production processes. It is frequently the case that the users of their products waive testing requirements in return for lower prices, and perform their own incoming inspection to whatever degree they find appropriate. In almost every situation in which extensive manual testing is used, the testing is performed under well-established procedures and data are carefully recorded.

With larger electronic systems, the products of a large number of vendors find their way into the completely assembled system. Very few electronic devices containing more than four or five printed circuit cards find their way to the end user without being involved in some kind of manual test procedure at some point during their manufacture. For this reason, there is a great deal of interest in testing procedure and reliability theory.

Manual methods are expensive on a per-unit basis, but in many instances they represent the only available means of testing a device properly. For this reason, manufacturers carefully study reliability and failure modes in their devices with a view to spending less time testing devices that are less likely to fail than other subassemblies.

No major manufacturer of components relies entirely on completely manual testing methods. For transistors and two-terminal passive devices, a number of automated machines are now available that speed up this process and make large-scale testing economically feasible. These machines rarely involve any significant degree of computerization. Generally, the components are mounted in some kind of carrier and fed to a test station where a simple, momentary test is performed. If the component fails the test, it is shunted into a reject bin. Counters keep track of the number of components accepted and rejected and these counts are recorded to become part of the manufacturer's reliability and production data.

Integrated circuits have placed an additional testing burden on the component manufacturer. Perhaps the greatest single contributor to this burden is the increased value of a single integrated component. Although the cost per function was reduced by integration, the dollar value of a single mechanical piece increased. To throw away an integrated circuit component is to throw away more dollars. The manufacturer is therefore motivated to test more thoroughly, rather than discard a batch in which too many samples failed.

The logical complexity of the individual device has also increased considerably. A resistor can be tested by applying a voltage and measuring the current. This is not the case with an IC. Now both electrical and logical characteristics of the device must be tested. Thus the component manufacturer finds himself facing the same problems that were faced by the manufacturer of printed circuit cards. He is now making devices to be handled as components which perform the functions that were handled by an entire printed circuit card. The advent of MSI and LSI will exacerbate this situation.

The increased functional complexity of the component has also increased the testing problem for the manufacturer of printed circuit cards. Now, he puts a larger number of functions on a single card—so many, in fact, that the use of general-purpose integrated circuit cards in large systems may be declining. It is difficult to make a general product and it is increasingly the case that printed circuits are being made to order. Manufacturers of minicomputers, for example, generally use large specially made boards. Many are putting the entire central processing unit on a single large board and the basic memory unit on another. These devices, although they are roughly the size of a few printed circuit boards, represent large-scale systems in the functional sense and are tested manually in their final operating situations.

Major semiconductor and printed circuit card manufacturers have started to use computerized testing stations that they have developed for their own in-house use. Smaller manufacturers still rely on spot checks, random sample tests, and general production controls to keep bad components out of their shipments. Large users who get their components from several sources have also developed in-house testing systems and devices so that they can cover any lapses on the part of their vendors.

While essentially manual methods are used for the testing of large systems, there are some devices that are available for testing fully and partially assembled systems. The new science of digital signal processing has produced several flexible machines for general spectral analysis. Punched-tape controlled machines are available that test wiring for continuity, shorts, leakage, and so on. There are many *ad hoc* testing devices developed for the checkout of particular devices and systems.

STOCKING AND DISTRIBUTION

With small manufacturers, a formal stocking and distribution system is not generally used. A stockroom serves to supply the manufacturing process with raw materials and a clerk watches for low stock and reorders. On the output side of the process, the low capital organization manufactures only when it has an order and does not have much trouble keeping track of what it has manufactured.

With larger manufacturers, and manufacturers of large systems, the problems are of different degrees. For the former, there is the matter of scale (large numbers of parts to keep track of). For the latter, there is the matter of complexity. Incoming raw materials and parts are obtained from a relatively large number of vendors. Receipts, on-hand quantities, and reordering must be carefully coordinated to keep the manufacturing process producing at the required rate. On the output side, the large manufacturer generally sells in a number of different markets and to a number of different customers. He frequently has more than one product.

For these reasons, the larger manufacturers in the electronics industry have adopted formal inventory and distribution systems which frequently use computers.

These systems fall into two major categories: manual systems with formalized procedures and computerized systems which are supported by formalized data collection and input/output procedures. These systems have formal administrative links with the actual manufacturing organization in a large company. The equipment used in these systems does not physically handle the output of the manufacturing process. That is, people move the products about, put them in warehouses, record the fact that they are there, and remove them to fill an order. Wherever a computer is used, it is largely used to analyze the data collected at various observation points within the system. The results of these analyses are delivered to management personnel who deliberate on them and either allow things to continue as they have been or institute changes. More details on the operation of these systems is given in Chapter 6, when we discuss the integration of the manufacturing process.

Chapter Four

The Modern Digital Computer

The purpose of this chapter is to familiarize the reader with the characteristics of modern digital computer systems and with some of their applications, so that he can be in a better position to evaluate the applicability of a digital computer to his own electronic manufacturing process.

We begin with the classical structure of the computer system and the principal types of computer systems and components. We then discuss general computer applications and treat four particularly interesting ones in some detail.

The discussion of applications is followed by a section on the subject of computer system architecture and programming languages. We relate this subject to the applications, so that the reader knows what computer systems features are needed for solution of particular application problems.

Finally, we discuss some of the general capabilities and limitations of computers, the problems and special considerations of using them, and some of the trends in computer development that strongly influence their utilization in computer-aided manufacturing.

We assume that the reader is familiar with some of the fundamental concepts of computer theory from his knowledge of electronics, namely, familiarity with nondecimal radix numbering schemes (such as binary and octal) and the general concept of storing instructions (programs) in the computer memory for sequential and conditional execution.

The reader may use the references listed in the Bibliography to amplify his knowledge of these areas, if he finds the need.

CLASSICAL STRUCTURE AND PRINCIPAL TYPES

A digital computer system is classically considered to be composed of three major subsystems:

1. A central processing unit (CPU)
2. A memory
3. Peripheral devices

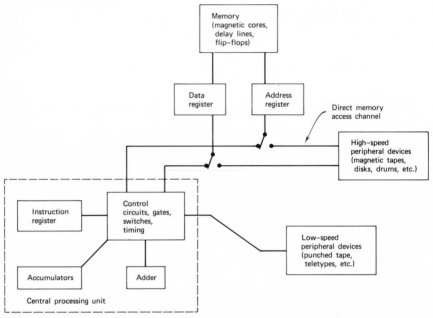

Figure 4.1. Classical structure of a digital computer system.

Figure 4.1 shows a classical computer system structure and some of the data paths through the subsystems.

The classical CPU consists of an adder, some registers having general holding, transfer, and shifting capabilities, and various control and status sensing logic. The CPU is the central control element of the system, executing the instructions specified by a program stored in the memory. We treat the internal structure of the CPU in detail when we discuss computer architecture later in this chapter.

The memory is usually organized into words that are groups of bits, generally of the same dimensions as the arithmetic registers in the CPU. The memory is most often fabricated from magnetic cores, although some older computers use delay lines. Programs and data in various stages of processing are retained in the memory.

The peripheral devices are used for input of data to the CPU and the memory, recording of the results of processing, and temporary retention of intermediate results. Typical computer peripherals are card and tape punches, magnetic tape drives, rotating memories (such as oxide-coated disks and drums), and printers. Computers must also communicate with machines, processes, and other computers. In several systems that we discuss, it may be difficult to determine what device is peripheral to what other device. However, the principles of interconnecting the computer to these devices and to other computers are frequently the same principles that are used in communicating with ordinary computer peripherals.

In many of the older digital computers, and in some of the minimal modern computer configurations, input and output data are transferred exclusively through the CPU. The CPU selects a particular peripheral device, obtains a word of input data from the device, and places this word in the memory. For output, the process is reversed. The word of data is transferred between the peripheral device and one of the CPU registers. During the transfer, the computer program devotes its full attention to determining that the device is ready for the transfer, transferring the word into or out of a CPU register, and reading or writing the word from or into the memory.

However, in many computers there is available either as an optional or standard item a direct data path between the memory and selected peripherals. This path is generally called a direct memory access (DMA) channel. Most computer memories have only a single "port"; that is, only one word can be stored or retrieved at a single time. For this reason, when the DMA is actively transferring a word it monopolizes access to the memory and prevents the CPU from retrieving its next instruction or data word until the DMA releases the memory port. Depending on the sophistication of the DMA hardware, the memory port may be monopo- lized for only a single cycle of the memory or it may be monopolized until the DMA has transferred a complete block of data. In the first case, the term used for the process is "cycle stealing." The second method is termed "data breaking" or "hogging (hog mode)."

Digital computer systems are divided into two principal types: business computers and scientific computers. In many cases one can perform the functions of the other, although less efficiently. There are many subdivisions of each class. Business computers are generally distinguished by the inclusion of two features:

1. A memory structure in which the "word length" is the number of bits required to represent a single character

2. Instructions for the handling of characters and the execution of deci- mal, as opposed to binary, arithmetic

Business applications generally involve sorting, merging, and retrieval from files of text as well as numbers. For this reason, the ability to directly address individual characters is quite useful. The introduction of the IBM System 360 modified somewhat the business computer's unique claim to character-oriented memories and several nonbusiness-oriented computers now have memories with this kind of organization. We discuss the motivations behind this organization when we discuss computer architecture.

There are two major reasons for providing decimal arithmetic capabilities. One reason is that decimal arithmetic makes the job of writing business programs easier. The other reason is that accounting applications require the precise representation of monetary amounts. Since there are terminating decimal fractions that do not terminate when represented in binary notation, monetary amounts cannot be represented with complete accuracy in binary form. Of course, special programs could be written that treat dollars and cents as separate quantities, relating carries and borrows between them. However, the general use of these programs would greatly increase the running time for problems. With decimal arithmetic, groups of four bits are used to represent a single-decimal digit. A carry from the digit is generated whenever an operation results in a binary value greater than 1001, the representation for nine in the decimal base. The carry/borrow operations are handled by hardware.

When arithmetic is required, as in accounting problems, it is as a rule relatively straightforward. The equations used are not complex and they do not require large numbers of iterations. The dynamic range of the numbers used is also relatively limited. Figures in excess of ten to one hundred million are rarely encountered. With scientific applications, extremely large and infinitesimally small numbers may be required. Thus scientific computers generally offer arithmetic instructions that process numbers in "floating point" notation. In floating point notation, numbers are represented by groups of bits divided into two fields. One field represents a power of two and the second field represents a binary fraction. The number is considered to be the product of the fraction and the number two raised to the power indicated in the other field. The fraction is termed the "mantissa," as with logarithms. The power of two field is termed the "exponent." Such a number representation has a much greater dynamic range but is not as accurate as the decimal or "fixed point" representations. However, since scientific computations do not have to account for every penny, they can trade precision for faster operating speeds and greater dynamic ranges.

The nature of scientific calculations is quite different from that of business calculations. Scientific calculations are typified by the solution of differential equations, matrix problems, and so forth. The equations are more complex than those used in business computations and are frequently cast in recursive forms that require many iterations. For these reasons, the speed of the central processing unit and the speed with which instructions and data can be retrieved from the memory are extremely important.

Both business and scientific computers offer hardware for fixed point arithmetic in which the number is represented in the computer in the ordinary binary sense. The hardware treats these numbers as integers and the programmer is responsible for keeping track of their scale—the position of the decimal or the binary point.

There are several digital computer models that aim at applications in both the scientific and business areas and, therefore, offer mixes of the features that we have just noted. The IBM System/360 is an example of such a computer.

Up to this point, we have discussed the principal features that distinguish two kinds of computers in the area of computation. However, modern data processing involves a great deal more than calculation. It involves the examination, modification, and rearrangement of text, as well. It involves the transfer of large amounts of this text into and out of the computer. In real-time and process control applications, which we will examine in some detail later, it involves the transfer of commands and data represented by binary words at the interface of the computer. It also involves the control of devices that act on and generate these words. Both business and scientific computers come with additional hardware features which help this kind of processing along.

Principally, these features are related to the input/output structure of the computer. We have already mentioned the direct memory access channel. Such a feature is clearly to be desired when large amounts of data are to be transferred. An additional feature is the "interrupt." In a computer equipped with interrupts, when a particular peripheral device has completed a commanded operation, it signals the central processing unit. This signal causes the computer program that is presently running to be interrupted temporarily. The mechanics of this process vary from one model to another, but generally it takes the form of causing the execution of an instruction stored at a particular, predefined, location within the computer memory. This instruction may initiate the execution of a separate computer program which takes whatever action is required by the condition that caused the device to send the original signal. This program, generally termed an "interrupt response program," notes the point at which the original program was interrupted and returns to that point when it has completed its own processing.

There are two principal advantages to having a computer equipped with interrupts. First, events occurring in the world external to the computer can be promptly detected and acted on. This does not mean very much to the business computer processing an accounting problem or to the scientific computer solving a set of differential equations, but it means a great deal to a computer involved in a process control application where a process alarm must be detected and corrected as quickly as possible.

The second advantage is that the data processing task can be run more efficiently, requiring less total computer time. If the computer is equipped with a DMA channel and interrupts, the computer programs can start an input/output operation and perform other processing while the input/output operation is going on. When an interrupt signals the completion of one input/output operation, a second operation can be started immediately. Since the transfer speed of most computer peripheral devices is much slower than the CPU/memory processing speed, the computer is placed in the position of waiting for a slower device much less often.

Having discussed the subject of computer types, we now consider the matter of "scale." Although scale is principally measured by the cost of the system and the additional features that these dollars buy, the delimiting points are very hazy. The lower end of the scale has recently been used as the entry point for what could be considered a new class of computer—the minicomputer. Further extensions of the scale to lower prices have recently brought about the entry of the programmed controller—a kind of little brother to the minicomputer.

The traditional approach to scale is quite simple and frequently imprecise. It works in the following way:

1. A large-scale computer system is very expensive.
2. A medium-scale computer system is moderately expensive.
3. A small-scale computer system is inexpensive.

Exactly how many dollars it takes to make a computer system expensive is quite difficult to determine precisely.

First of all, it is very difficult to translate such terms as computing power, "throughput," and so forth, into a figure that represents the cost of a computer system offering these advantages in particular quantities. For one thing, the kind of problem to be solved by the computer heavily influences the dollars that need to be invested to obtain a computer system with the requisite capabilities. If the problem requires a great deal of calculation, a fast central processing unit gives good return on investment. However, if the problem involves a great deal of input and output and few calculations, then the money is better invested in fast peripheral devices.

H. R. Grosch of the National Bureau of Standards has speculated that the "power" of a computer is proportional to the square of the cost of the system. If computer system *A* costs twice as much as computer system *B,* then computer system *A* is four times more powerful than computer system *B.* This rule is known in the computing trade as Grosch's law. Those who have found it to be true for their particular kind of problem treat it as gospel. Those who have not treat it with skepticism. In general, when dealing with computer systems, it is best to analyze the problem carefully rather than trying to rely on rules of thumb and folk lore.

Computer system dollars can be spent in any one of the three major parts of the computer system: CPU, memory, or peripheral devices. In the areas of memory and peripheral devices, it is generally spent on extra capacity or speed. In the CPU area, it can buy both faster processing (although generally the memory speed is the limiting factor here), a larger set of instructions, or more working registers. With a more powerful set of instructions, more work can be done in fewer instructions. This has the effect of cutting down on memory requirements and improving processing speed. A larger number of working registers in the CPU, where intermediate results can be temporarily retained, means fewer accesses to the memory. This also improves processing speed. In some computers, these extra registers can serve the function of "index registers." These are independent pointers to memory locations and counters. Their presence frequently simplifies the programming task and reduces the number of instructions required to carry out a particular function.

Extra money spent on memory can mean either a faster memory, a larger one, or a combination of both. Although there are occasionally unusual exceptions, it is generally the case that the computer memory must be large enough to contain the complete computer program. With most computer systems, the capacity of the memory can be expanded without changing to a different model. The speed of the memory is a different matter. Since most CPUs are synchronized with the speed of the computer memory, utilizing a faster memory usually means replacing the CPU with a different model, as well.

In the area of peripheral devices, there is a great variety from which to choose. If the application problem requires access to a large amount of data in a way which is not sequential, then a disk or drum memory may be required. Large data files that can be processed in a sequential manner are generally written on magnetic tape. If a large amount of human-readable output is required, then a line printer is generally needed. Recent developments in the area of computer peripherals include inexpensive disk memories, low-cost line printers, and graphic terminals for use with minicomputers.

The minicomputer is a very recent phenomenon in the computer market. The first commercially successful minicomputer was the PDP-8, manufactured by the Digital Equipment Corporation. Since the inception of the PDP-8, many computer manufacturers have adopted a minicomputer line and many companies have been formed for the sole purpose of manufacturing and selling computers in this class. At the present time there are over thirty corporations, offering over fifty different models, engaged in the manufacture of minicomputers. With respect to sophisticated design, general utility, and speed, many of these minicomputers compare favorably with more expensive models and some of them exceed the performance of many computer that were in use during the late 1950s and early 1960s. Their architectural features are, in many instances, remarkably similar to central processing units costing considerably more.

The usual definition of the minicomputer is based on the cost of the central processor, not its physical size. Some define a minicomputer as having a central processing unit that costs less than $20,000. Others define it as having a minimum useable configuration costing less than $50,000. Although it should be obvious, we point out that a central processor by itself is rarely of any practical use. Generally, there must be some kind of peripheral device with the system.

If the computer is to be used for calculations or general data processing, some kind of output medium must be provided—a printer or punch, for example. If the computer is being used in a control application, there must be some means of communicating with the devices being controlled. It is often the case, especially for the applications in which we are most interested, that calculating the exact cost of the required computer and interface/peripheral devices is a far from simple matter.

Almost all of the fifty minicomputer models are now selling for about $10,000. This is the cost of the CPU alone, although in several instances the price includes the cost of an ASR-33 Teletypewriter. Many of the standard peripheral devices used with these minicomputers actually cost more than the CPU. A high-speed magnetic tape, for example, with moderate bit-packing densities and tape speeds, costs between $15,000 and $20,000. A line printer of reasonably good speed costs nearly $10,000.

Historically speaking, $10,000 for a computer is a very small price indeed. These relatively low prices have produced several very interesting effects which we might digress for a moment to consider.

Perhaps the most notable effect of this low price has been a run on the minicomputer market. At the present time, many of the minicomputer manufacturers are doing reasonably well considering level and amount of competition. Many possible users, attracted by the low price, are now considering these computers for certain applications. Generally, the cost of the application is somewhat more than they may have expected when they first began looking into the matter. However, once their attention is attracted they rarely back off from finally going ahead and using the computer.

The minicomputer price war has also caused pressure on the minicomputer manufacturer to lower his prices even more, either through direct price cuts on existing models or through the development of new models which are less expensive. One outcome of this has been the introduction of what has come to be called the "programmed controller." Several of the present models of minicomputers, when viewed through the jaundiced eye of the professional programmer, are more properly termed "controller" than "computer" because of their extremely simple architecture and the limited computing and processing power and speed that they offer. In fact, the names given to them by their manufacturers often support this viewpoint. The PDP in the PDP-8s name stands for "programmed data processor," a beginning of this concept. Another line of computers uses the initials and name of "stored program controller." These names were originally chosen not so much for accuracy as to overcome fears on the part of potential customers over buying a computer. Before the minicomputer, most large organizations used computers entirely for management and financial data processing. As a result of this, the procurement of computers became the official bailiwick of upper-level management staffs, largely operated by personnel who knew only business data processing. These people were not qualified technically to evaluate a computer for use in a process-control application, and those directly responsible for the applications simply did not want to take the trouble to work with them. The most expedient way around this problem was to avoid calling the piece of hardware to be purchased a computer.

The price squeeze caused by price reductions also leads to other effects. For example, there simply is not very much room left in which to reduce prices. A ten or twenty percent reduction now represents fewer dollars and is becoming less important for many applications. Some of the manufacturers are countering this effect by offering more powerful computers for the same price—a kind of lateral movement. One of the minicomputers that presently sells for slightly more than $10,000 is offering architecture that is very similar to that of the large IBM System/360, complete with a subset of the instructions for that computer. Since the manufacturer of a computer cannot patent the format and results of a computer instruction, some manufacturers are now building computers that have the same instruction set as a competing computer. Having to invest less money in general design and software, and taking advantage of technologies that may not have been available when the competing computer was designed, these manufacturers are able to offer a virtually equivalent computer for less money.

The high cost of peripheral devices relative to CPUs, which we have already noted, represents an inversion of the intuitive values of many people who deal with these machines. It does not seem right that the peripheral devices, even a single one, should cost more than the central processing unit. The response of the industry has been to attend to the development of low-cost peripheral devices specially tailored for the minicomputer system. As a result, peripheral prices are now dropping drastically. They have to catch up with the still dropping prices of the central processing units, but they are trying and may achieve some success in the next few years. At the present time, the greatest expectations for reductions in overall system prices rest on the reduction of peripheral device costs.

Now that we have looked into the various classes of computers, we will devote some attention to their applications, so that we can get a better idea of how they can be used in computer-aided manufacturing systems.

APPLICATIONS

It has been said that the potential applications for the digital computer are unlimited. This may be so, but it is also true that the potential applications of the digital computer are not all-encompassing. There are some things that a computer cannot do well or economically. There are other things that a computer cannot do at all. About all one can say with certainty is that the set of computer applications exhibits highly dynamic behavior—it is changing and expanding continually. The principal driving forces are technological innovations and economic developments.

To put some of the foregoing material into perspective and to set the stage for discussions on architecture, languages, and applications in computer-aided manufacturing, we are now going to discuss some applications that represent typical uses of the digital computer.

Each year "Computers and Automation" magazine publishes a computer directory and buyers' guide. The 1969 issue of this guide lists over 1600 applications of digital computers, classified by general field. Clearly, we cannot begin to treat them all. Instead we will cover four classes of applications that are pertinent to the subject of computer-aided manufacturing. We treat these application classes in a general way so that the principles stand out, uncluttered by details.

There are two major computer applications that we will defer to the next chapter: automated materials control systems and direct numerical control systems. These are the two major computer applications in manufacturing, and we will treat them in detail in Chapter 5, borrowing from the principles that we learn in this chapter.

The four application classes that we cover in this chapter are the following:

1. Stand-alone systems
2. Process control systems
3. Tracking systems
4. Time-sharing systems

The set of digital computer applications can be divided into two parts, consisting of "real-time" and "nonreal-time" applications. Of our four examples, the first is a nonreal-time application. The others are real-time applications.

The concept of real time is somewhat difficult to define in a precise way. It is one of those concepts that we so often find which everyone understands well enough to recognize an example when they see it, but not well enough to define precisely in the abstract. The earliest attempts to define real time used the idea of time constraints—if the system had to do its work within a certain time, then it was a real-time system. The problem with this definition was that it could be applied to define as real-time systems some systems which did not match the intuitive concept of real time. For example, the preparation of payrolls has never been historically considered a real-time problem. However, payrolls must be met precisely, and in this sense they meet the earlier definition. The sort of system that *does* match our intuition is a radar tracking system or a process control system.

We are not going to try to resolve this issue in this chapter. We generally consider a system to be real time if its response must be delivered to the user of the response within a few seconds from the time that the system was stimulated. We also require that the system response be delivered to a device which is not a normal data processing peripheral. For example, a machine tool or a time-sharing terminal.

Stand-Alone Systems

Stand-alone systems represent the most prevalent digital computer system. A typical installation is shown in Figure 4.2. These systems are configured with standard computer peripherals in varying combinations: magnetic tape drives, disk and/or drum memory units, card handling equipment, line printers, and a control console. The minimum useable configuration is the central processor, the memory, and a control console which includes some kind of keyboard/printer device, such as a Teletype unit or electric typewriter.

These systems are typified by the large- and medium-scale business and scientific data processing systems manufactured by IBM, Univac, Burroughs, Control Data, and Honeywell. The largest portion of the nationwide investment in computers is in this kind of system, which can be found in almost every corporation with annual sales over a few million dollars.

In some cases, inexpensive minicomputer systems are being used in stand-alone configurations for particular segments of the work normally handled by large- and medium-scale stand-alone systems. Software packages for

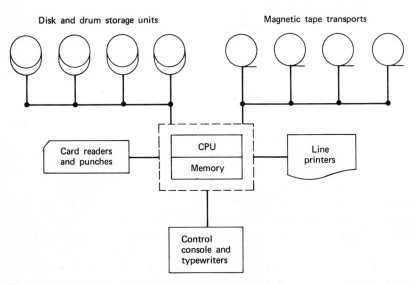

Figure 4.2.   Typical stand-alone computer system.

payroll, accounting, and inventory control are developed and sold along with the computer system for modest sums and the computer is then used almost exclusively for the particular application segment. Companies that develop these systems frequently buy the computer from the original manufacturer and repackage it for use in an office or storeroom environment.

Listed below are some of the typical uses of stand-alone computer systems:

Business and Production
    Accounting—preparation of general ledgers, statements of income
    Operations research—study of various operating plans, market analysis, economic product mixes
    Payroll—preparation of paychecks, calculation of taxes, payroll summaries

Investment analysis—balance sheet analysis, investment simulation, market history analysis

Inventory control—maintenance of inventory counts, automatic re-ordering

Production control—work-in-progress analysis, preparation of work sheets, parts explosion, purchasing

Numerically controlled programming—parts programming

Scientific and Engineering

Simulation and analysis—studies of control systems and physical phenomena, operator training, and so on

Trajectory and orbit analysis—planning of space missions, satellite orbit life

Architectural and civil engineering—structural analysis, heating and lighting, materials costs, land fill and excavation volumes

Circuit and system analysis—ECAP, CIRCAL language analysis of circuits, simulation of digital and electronic systems, time and frequency domain analysis

Meteorology—weather prediction, hurricane analysis

Numerical analysis—solution of differential and integral equations of physics and chemistry

Process Control Systems

Process control systems represent a major example of real-time digital computer systems and employ many of the techniques that are utilized in computer-aided manufacturing systems. They have been principally used by the large heavy industries such as automobile manufacturing, petroleum processing, and steelmaking.

Figure 4.3 illustrates the structure and components of a typical process control system in which the digital computer is the central control and monitoring element. The complexity of a process control system is generally measured in terms of the number of "loops" that it controls. By a loop we mean a separately controlled production element that involves feedback either through the computer or through an external controller for which the set point has been specified by the computer.

The principal inputs to the computer are derived from electrical and mechanical devices whose states are altered by events or changes within the process itself. These include analog sensors (such as current-measuring devices), transducers that convert capacitance, physical motion, temperature, and so on, into analog electrical signals, digital sensors and contacts which convert discrete events into direct current levels or pulses, and

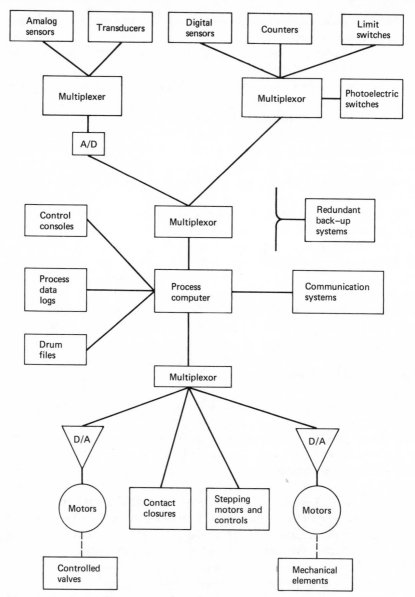

Figure 4.3. Structure and components of a typical process control system.

counters that may be used to accumulate the number of times that these events occurred within a specific time frame. Since the number of input channels into the computer is generally limited to some small power of two by the input/output addressing structure, and because there are frequently many more devices to be monitored and serviced than are represented by this number, the various inputs are usually grouped and selected by one or more multiplexing devices which then transfer the data directly into the computer. Since analog-to-digital (A/D) converters are relatively expensive, analog signals are also multiplexed into one of a few A/D converters in the system.

Outputs from the computer system are selected by the computer programs on the basis of programmed control rules whose parameters are the inputs to the system in their various forms. These outputs may take the form of motor speed commands (converted to analog forms that specify current application to the motors), energization of relays to provide contact closures, pulses, and counter settings which drive synchronous and stepping motors. The contact closures usually cause a single event or a programmed train of events to take place, or they energize a particular device for continuous operation. The motor outputs are generally used to open and close valves or to cause the physical motion of a device or piece of equipment. As is the case with inputs, most outputs are multiplexed.

The process computer system also reports the state of the process on various special display panels, so that human operators know what is going on. Most of these consoles provide controls so that operators can take over manual control of the process in the event of malfunctions or special conditions. Supplementing these control consoles are data-logging printers on which the process computer records a log of events in the system, materials used, and so forth. Large drum files are often used to buffer data and to hold the data files used by the process computer in determining control settings. In some cases, these drum files are also used to retain programs which are read into the core memory when they are actually needed at some point in the control procedure. As we see when we discuss input/output devices later in this chapter, drum files provide quicker access to data and better reliability than most other standard high-speed computer peripheral devices. Data are recorded on drums in tracks that move under read heads as a result of the rotation of the drum. One read head is provided for each track. Disk memories are also available that provide one read head per track and are frequently used in the place of a drum file.

The process computer may also be required to communicate process status and to control equipment at remote stations. Thus when distances greater than 2000 to 3000 feet are involved, a special communication system is used which involves the transmission of information and encoded control signals over public or private telephone communication lines.

With large- and medium-scale process control problems, the process is generally highly dependent on the continued functioning of the process computer system. Interruptions in the continuity of the control are not tolerable and manual back-up controls may not be able to handle the process for the amount of time that it may take to repair the malfunction that caused the interruption. For this reason, a back-up computer system

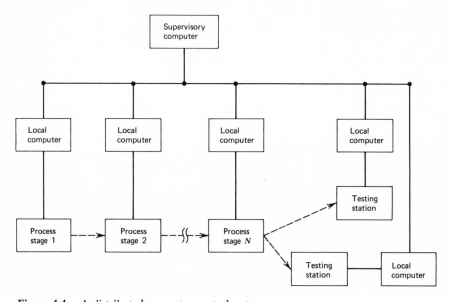

Figure 4.4. A distributed computer control system.

is generally included on a continuing standby basis. If a malfunction in the main process control computer system is detected, the back-up system is switched on-line (either manually or by its own initiative) until the main system has been repaired and restored to normal operation.

The system depicted in Figure 4.3 is an example of a classical process computer system. The advent of the minicomputer, however, has brought about some interesting changes in the structure of more modern process computer systems. Primarily, it has made the "distributed computer system" economically feasible. Figure 4.4 illustrates the concept of such a system. With this kind of system structure, local control is placed in the hands of a local computer that is responsible for the direct control of only a small portion of the process. The local computer handles its control function using data and commands supplied to it by a computer at a higher, supervisory level. This supervisory computer coordinates the activities of a number of local minicomputers and is not concerned with the millisecond by millisecond direct control of system elements. If the process control system is a particularly large one, we may find the supervisory computers working from directions provided by a still higher level of computers.

Having separated and distributed the control elements in this way, the process system gains in reliability and immunity to malfunction. If one of the minicomputer controllers breaks down, only its section of the process is affected. The supervisory computer can detect and report the malfunction promptly and the minicomputer can be literally replaced with a spare which is kept on hand for just this purpose. Thus down time is kept to a minimum. If a supervisory computer malfunctions, it can be similarly replaced if this is a cost-effective maintenance plan for the particular application. In any case, its malfunction would not prohibit the minicomputer controllers from proceeding with local control of the process. They would be held up only when they "ran out of work" and had to await new directions from the supervisory computer. If the supervisory computer were not ready, the local computer could hold the process or close it down in an orderly manner ("fail soft") until directions became available.

The initial installation of a process control system and field modifications to the system represent major economic and schedule problems. Distributed systems also offer obvious advantages in these areas. In installing the system, the local controllers can be installed and tested first. Then, the remainder of the system can be built on them. If problems arise, they can be more easily traced, generally to the newer elements of the system. In making system modifications, usually only the local control areas are involved. This means that software and hardware modifications are localized. In some applications, a new element can be installed and tested in parallel with the unmodified process while the unmodified process is in full operation. The new element can then be switched into the process when it is fully tested.

Tracking Systems

As our second example of real-time computer systems, we have selected tracking systems, such as the one depicted in Figure 4.5. These systems exhibit many of the same features and problems that are found in process control systems, but frequently in more severe and constraining contexts.

The system shown in Figure 4.5 is a steerable disk tracking system such as those used in instrumentation radars. The control system has essentially three loops: the estimation of range and the control of a radar "range gate," the estimation of azimuth and the control of the azimuthal pointing direction of the antenna, and the estimation of elevation and the control of the elevation of the antenna. The range of the target is measured by the delay between the transmission of a radio frequency impulse and a returning reflected wave. The angular position of the target with respect to the angular position of the antenna is obtained from phase relationships

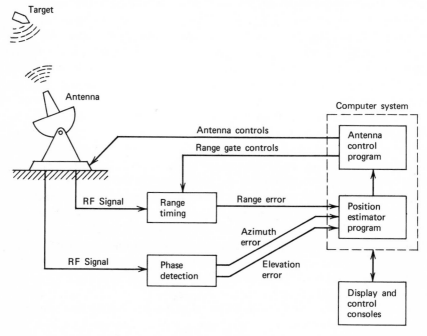

Figure 4.5. Computer controlled tracking system.

in the returning wavefront. These measurements represent measured errors in the estimated position of the target and are used to upgrade the estimate maintained by the computer. The revised estimate is then converted into pointing directions that steer the antenna.

The cycle time for this control loop (measure, revise estimate, redirect antenna) must be short enough to maintain a smooth and accurate track of the target. This time is a function of the accuracy requirements of the system and the dynamics of the target vehicle. A reentry vehicle, for example, presents a much more difficult tracking problem than an aircraft or a stellar object. Generally, these cycles run from 10 to 50 milliseconds and, more rarely, up to 1 second. The calculations involved in maintaining the track and in converting the target position to a form useful for ancillary systems and displays are quite complex. It is usually the case, therefore, that the computer is strongly dedicated to this single task. We rarely find computers directing a number of antennas for which the tracking loops are closed through the computer.

There are many other radar tracking configurations worthy of mention. Particularly prominent is the phased array radar, which can be classified as an electronically steerable antenna. With a phased array radar, it is possible to track many targets since the antenna can be repositioned at electronic speeds to collect position data first on one target, then on another. A second prominent type is the scanning radar, used for detection, warning, and surveillance. In this system, the antenna is continually rotated mechanically, illuminating sections of its environment as it swings by and detecting reflected returns.

The bearing that computerized radar systems have on the subject of computer-aided manufacturing is largely historical. They represent some of the earliest sophisticated real-time systems, and much of what we know about real-time techniques was developed for use in these tracking systems. The need for recursive estimation and control algorithms led to the development of many mathematical programming techniques now employed in process control systems for optimization. The need for real-time software operating systems formed the basis for the time-sharing systems and process control software systems that are now in widespread use.

Time-Sharing Systems

In Chapter 3 we described time sharing as a method of using a computer in which the individual users are given access to the computer in a way which makes it seem that they have continuous use. A time-sharing computer system involves the use of special hardware and programs to accomplish this purpose. The structure of a typical time-sharing system is shown in Figure 4.6.

There are two major problems to be treated in the design of a time-sharing system:

1. The problem of communication between the user and the computer system

2. The problem of running the user's programs, retaining them, protecting them, and providing all the users with fair access to the resources of the system

These problems can be solved with a combination of specially designed hardware and software.

The system consists of the following major elements:

1. Terminals at which the user may enter programs and commands to the system and receive the output of processing performed for him by his programs and the system

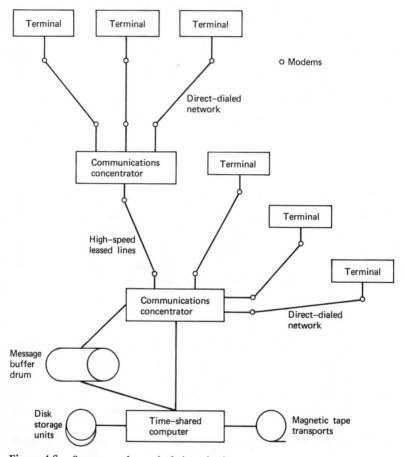

Figure 4.6.   Structure of a typical time-sharing system.

2. Communication lines, frequently the direct-dialed network of the public telephone system

3. Data concentrators, which collect transmissions from the slow-speed telephone lines and transmit them either directly into the time-shared computer system or to the system over higher-speed lines

4. A time-shared computer system on which the user's programs are run

The most common time-sharing terminal is the Teletype Model 33 ASR (automatic send-receive). This device uses an eight-level ASCII code for information transmission and transmits at a baud rate (bits per second) of 110. It includes a keyboard with regular typewriter symbols and special control functions and symbols, a printer (character serial), a paper-tape perforator, and a perforated-tape reader. The basic character rate into and out of these terminals is ten characters per second.

Graphic terminal devices are also available for use in place of the teletypewriter. These devices are frequently plug-compatible with channels designed for use with teletypewriters. They display transmitted and typed characters on a CRT screen and are generally operable at three to ten times the speed of the teletypewriter. They suffer from a lack of hard copy capability, but in some cases can be used to drive a teleprinter to get around this difficulty.

Connection with the communication network can be effected in several ways. Semipermanent terminal installations are generally used in which the terminal is connected to a data set provided by the telephone utility. Acoustic coupling devices are also available that allow the terminal to be connected to the communication network acoustically through an ordinary telephone handset.

Most time-sharing systems employ a data concentrator at the "front end" of the computer system. This device collects messages from the communication network at their transmission speed of 110 baud and passes them on to the computer system or places them in a message buffer (often a high-speed drum) from which they are selected by the time-sharing computer. When the system is widely spread geographically, the data concentrators are used to collect information locally and forward it over higher-speed lines. This technique reduces costs due to line charges from the telephone utilities involved. The concentrators are also used to distribute replies from the computer to the user's terminal by reversing the above process. These data concentrators are either hardwired devices or, in many cases, specially programmed minicomputers.

The computer is usually an example of a large-scale computer system. It is equipped with a large-capacity core memory and high-speed random access bulk memories such as disks and drums. High-speed magnetic tapes are also employed, primarily to retain back-up copies of the data and program files that are kept on the disk and drum memories.

Although the communication system and the speed and memory resources of the computer system are highly important, the key to a time-sharing system is its software. There are two main components to this software system: the operating system software and the application programs.

We have already touched on examples of these application programs in our discussion of some of the circuit analysis programs in Chapter 3. The application programs represent a collection of already developed computer programs designed to solve particular classes of problems on a time-sharing system. In general, the applications that they include fairly well cover the list of applications that we noted as being run on stand-alone computer systems. Accounting packages, investment and bond analysis, text editing programs, and numerical analysis programs are generally offered. Most time-sharing systems offer simple languages for the programming of straightforward calculation problems and the simulation of desk calculation machines. They also offer compilers for the FORTRAN, BASIC, and Algol languages. In most cases, the user writes his own programs using one of these languages. Some time-sharing systems offer programs that simulate minicomputers, so that users can assemble and test their minicomputer programs using the facilities of the time-sharing system. The text editing programs are frequently used to prepare and edit documentation, produce form letters, and so on.

The operating system software is responsible for communicating with the user via the data concentrators and communication network, running the user's programs and application programs in a controlled environment, and managing the distribution of the system's resources to the various users.

The two most important system resources are central processor time and core memory. There are many ways for the operating system software to distribute these resources, and there is still a great deal of research being devoted to the study of this subject and to improvements on present methods. We consider a simple example of one way in which these resources can be distributed as an illustration of the general principles involved.

In our discussion of interrupts earlier in this chapter, we described the way in which a peripheral device could signal the central processor that it had completed a particular task, such as the transfer of a block of data to or from the core memory. It is also possible to obtain an interrupt signal from a time-base generation device such as a crystal clock or a clock-driven counter. Such a method is used in time-sharing systems to divide real time into a number of sequential segments.

Suppose that several user programs are stored in the core memory of the computer. At the start of our example program A is running. (Refer to Figure 4.7.) After program A has been running for some period of time, generally 20 to 100 milliseconds, the time-base generator signals the central processor and causes an interrupt, which brings the operating system programs into play. The operating system programs then examine a table stored in the core memory, which lists the other programs that are in core. It discovers that program B is next in line in the queue that this

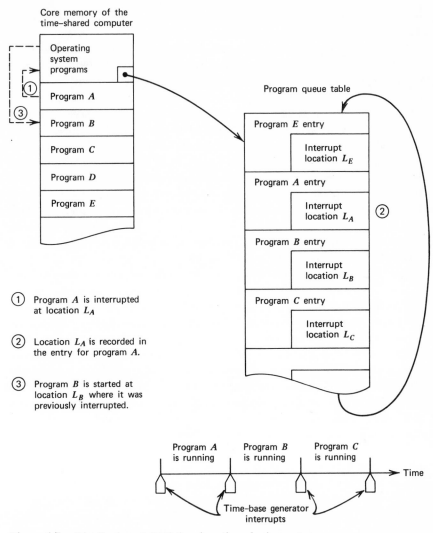

Core memory of the time-shared computer

Operating system programs

Program A

Program B

Program C

Program D

Program E

Program queue table

Program E entry

Interrupt location L_E

Program A entry

Interrupt location L_A

Program B entry

Interrupt location L_B

Program C entry

Interrupt location L_C

① Program A is interrupted at location L_A

② Location L_A is recorded in the entry for program A.

③ Program B is started at location L_B where it was previously interrupted.

Program A is running

Program B is running

Program C is running

Time

Time-base generator interrupts

Figure 4.7. Distribution of CPU time in a time-sharing system.

table represents. The operating system programs then transfer control of the central processor to program B. Program B was most likely interrupted earlier in the same way that program A was just interrupted. The table used by the operating system has an entry that records the point at which program B was interrupted, and it is at this point that the operating system restarts program B. Before restarting program B, however, the operating system records, in the program A entry, the point at which program A was interrupted. Thus program A may be restarted when its turn comes around again.

By slicing up real time in this way, the operating system can allow the user's programs to operate in a round robin fashion, with each user being given an equal "cut" of the central processor running time.

Things have been fairly simple up to this point. But now we must consider what happens if one of the programs requires more core storage than may be presently available. If this is the case, the operating system can look to see which program is farthest down in the queue and preempt its core space. The contents of this core space is generally written onto a high-speed drum from which it can be retrieved when its turn comes up again. If this does not provide enough core space, the next program's space is taken in the same manner, and so on until the core requirements are met. This process is called "swapping" and the user of the time-sharing system can generally tell when his program has been "swapped" out of core by an unusually long pause in the output of his program or an unusually long response time to a command that he has entered. Usually, these times do not exceed one or two seconds and rarely represent any inconvenience on the part of the user.

All input and output operations that are initiated by the user's programs are routed through the operating system software. When the operating system detects that the user wants to perform an input/output operation, it takes over control and starts the operation. It then passes central processor control over to the next program in the queue, so that the system will not be tied up while waiting for the input/output operation to be completed.

A recent count of time-sharing services revealed roughly ninety-five separate corporations offering these services both locally and on a nationwide basis. There are many people using time sharing for various purposes and the number is bound to grow. The economic aspects are generally very attractive for the user of these systems. He is given access to a very powerful computing and data processing tool for much less investment than would be required if he were to purchase his own system. His investment is less because he is sharing it with the other users of the system.

However, many of the vendors of time-sharing services are in financial trouble or are failing to realize the profits that they originally expected. This is not from lack of market response. It is due to the large costs of operating these systems and of maintaining a continuing lead in the technological development that is required to keep ahead of the competition. The time-sharing user is very fickle. If a service other than the one he is presently using begins to offer an application program that he finds attractive, he may switch over—giving the original vendor only 30 days' notice (time during which he does not have to actually use the original system). Vendors cannot tie down users with long-term agreements because no vendor has a sufficiently strong technical edge to compensate the user for the assurance that the user will stay with the service.

COMPUTER SYSTEM ARCHITECTURE

Computer system architecture is the science of designing computer systems. Like the words "physics" and "mechanics," it is also used as a synonym for the "organization" of a system. In this section, we examine some examples of computer architecture.

Computer system architecture has been in existence since the beginning of automatic computation, but it has only recently been granted the honor of a formal name and the beginnings of a formalized discipline. This dates most notably from the early 1960s and the introduction of the IBM System/360. A good deal of theoretical attention is now paid to this subject, both in the halls of academe and in the R&D departments of computer manufacturers. It has resulted in accelerated improvement in the quality and capabilities of modern computers.

There are six general categories of the subject of computer architecture to which we address ourselves. They by no means represent an exhaustive treatment of the subject. A brief glance at the table of contents of the *IEEE Transactions on Computers* will demonstrate the great variety of arcane topics that fall within the scope of computer system architecture as a scientific discipline.

The categories to which we address ourselves are the following:

1. Classical architecture
2. Microprogrammed computers
3. Decentralized structures
4. Input/output structures
5. Memory organizations
6. Multi- and parallel processors

We have already said a good deal about classical architecture in the early part of this chapter, and the reader should have a resonably good grasp of what is meant by a "classical architecture." However, we will have a few more words to say about this.

The next two categories illustrate computer architectures that are representative of many computer systems presently being offered in the marketplace. They are well-known, widely accepted design techniques, yet they also represent the very latest in excellent approaches to the design of a computer system.

The category of input/output structures is included so that the reader can become familiar with some of the details of the way in which the computer communicates with peripheral devices and the machinery that it controls.

The last two categories are included to familiarize the reader with some of the more advanced topics being treated by the science of computer architecture. Here, we discuss two of the advanced forms of memory organizations, other than the traditional single-port synchronized core memory. We also discuss the advanced processor organizations typified by the newer "supercomputers."

Classical Architecture

In the early parts of this chapter we discussed the relationship between the memory, the central processing unit, and the peripheral devices of a computer with classical architecture. We also mentioned some of the features of the central processing units such as interrupts, index registers, and so forth. We now investigate some of the features that accompany these—things that aid the computer programmer in his task of producing programs for the machine.

One of the first considerations for the programmer is the instruction set with which the computer is equipped and the way in which it addresses its data. We noted earlier that the memory of the computer is divided logically into groups of bits called words and that the number of bits in these groups is related in some way to the size of the data paths and the organizations of the central processing unit. In some modern computers, the bit lengths of the instruction words may be integral multiples of the memory word size, with different instructions having different lengths. In classically structured machines of the older variety, the instruction size is generally of the same dimension as the memory word. That is, one memory word contains one instruction.

The central processing unit generally operates by retrieving sequential words from the memory. A certain field, or group of bits within the word, is then logically decoded within the CPU to determine the actual instruction. If the number of bits in this field is c, then the number of possible logical states that the field can represent is 2^c, and this is the number of instructions that the field can define. This is the most straightforward way of defining instructions.

An instruction is an order to the central processor to do something. This order may or may not involve an "operand," an item of data to be acted on, altered, or used in some way. The instruction must specify this operand in some fashion, for example, define its location. The operand may be contained in a location within the core memory or it may be in a CPU register. The operand may even be related to an input/output device serving as the source or destination of the item of data. For these situations, three cases arise:

1. The operand is in a memory location. In this case, the instruction must specify an address within the memory.

2. The operand is in a CPU register. In this case, either the register is defined by the instruction itself (e.g., the instruction states that something in the A accumulator is to be shifted to the right or left) or a field in the instruction contains a code number that identifies the register.

3. If the operand location is an input/output device, a field in the instruction identifies the device by a particular code number.

We can see that there are also some instructions that do not necessarily involve an operand, for example, an instruction that causes the computer to halt—to cease executing a program. In this case, and in the second and third cases noted above, we can see that the full number of logical states of the instruction word may not be fully utilized since the number of bits needed to specify the operand may be small. For example, consider the case in which the instruction orders the computer to shift the contents of the A accumulator to the left a certain number of places. Is this shift to be cyclic? Is it to carry around to another register within the central processing unit? Having used the principal code field within the instruction word to identify the instruction as a shift instruction, can we now place a different interpretation on other bits in the instruction word and thereby provide a more detailed instruction to the computer—that is, elucidate on the instruction using bits left over from the smaller operand location code? The answer is yes, and this technique is generally employed to expand the instruction set of the computer.

Some of the features that we have just been discussing are illustrated in Figure 4.8. We have just given a hint that the computer designer is always struggling to get as much out of a computer instruction word as he can. Why? Because computer memory is very expensive relative to the cost of other central processor components. The smaller the computer word, the less costly the computer. The more work that can be obtained from a single instruction, the fewer instructions needed and, therefore, the smaller the memory needed. However, the smaller the word, the fewer the number of instructions and the less powerful the computer.

Let us look at another aspect of structuring the instruction word for a computer. Consider the case where the operand for the instruction is in a memory location. The instruction can only address as many core memory words as there are logical states in the address field of the instruc-

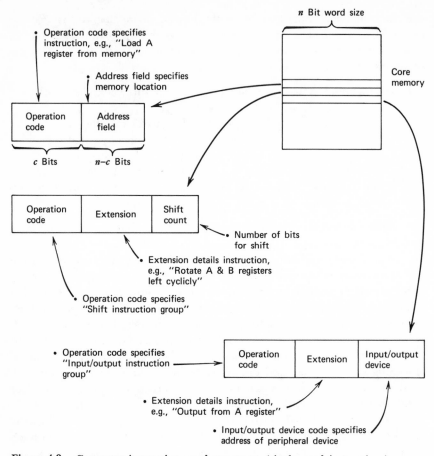

Figure 4.8. Computer instruction word structures (single-word instructions).

tion word. A core memory is generally configured in minimum quantities of 4096 words (a convenient power of two). In classical architectures, the memory can generally be expanded in multiples of 4096 words to some upper limit dictated by logical considerations and the addressing equipment associated with the core memory itself. To address 4096 words requires an address field that has 4096 logical states and therefore consists of 12 bits. The number of directly addressable words doubles as each new bit is added to the address field. Most of the classical computers employ from 2^{12} (4096) core memory words up to 2^{15} (32768) words, or more. Adding more bits to the instruction word so the computer will have a healthy complement of instructions generally makes the instruction word size from 24 to 36 bits and implies rather hefty memory costs. This approach has changed rather drastically in the more modern digital computers and particularly in the minicomputers in which we have a special interest.

Most modern minicomputers have memory word sizes from 8 to 16 bits. In the 8-bit computers, virtually no room is available within a single word for a complete instruction field comprehensive enough to provide adequate capability for general applications. Thus the instruction word size is generally taken as an integral of the 8-bit memory word size. The first 8-bit word may identify the instruction type and, if a longer word is required to fully specify the instruction, the succeeding one or two 8-bit "bytes" are used to fill out the instruction.

This approach is also used, on occasion, in minicomputers having 16-bit word lengths. The first word specifies the instruction type and other information such as shift counts, register selection, or input/output device number. If the instruction references an operand in the core memory, the second word can be used in its entirety to specify the address of the core memory operand. This multiword instruction approach is illustrated in Figure 4.9.

However, in most small-word-size computers, the technique of "paging" is employed. There are several techniques of paging, but in general they all involve dividing the memory into groups of consecutive words called pages. The number of words in these pages is a power of two, the power being equal to the number of bits left over for an address field after the instruction code fields have been specified. The instruction is then limited to accessing *directly* only the number of words that can be specified by this field. Since this number is less than the total number of words generally desired and actually contained in the core memory, the designer is faced with the problem of providing the programmer with access to the other words. This problem is solved by providing an indirect addressing mode and a control bit within the instruction that enables the programmer to select this mode for the duration of the execution of this instruction. In this indirect addressing mode, the operand location specified in the instruction refers not to the address of the operand itself but to a location that contains the address of the operand. More bits are available in this intermediate word and the address can be fully specified therein. The CPU uses the contents of this last location as the actual address for the operand.

The concept of paging is one of the major contributions of programming theory to the science of computer system architecture. Considering many different programs in a statistical sense, it was noticed that there is a tendency on the part of a program to access memory words that appeared principally within only two regions. One of these regions is the vicinity of the instruction. Instructions frequently refer to constants, programmed counters, and data that are "near" the instruction. That is, their core memory addresses are numerically close. The other region is a collection

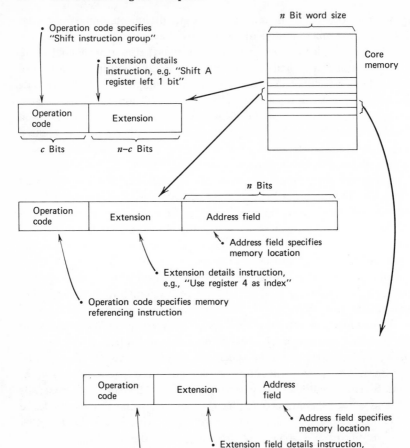

Figure 4.9. Computer instruction word structures (multiword instructions).

of locations that contain the data common to all of the subprograms that make up the complete collection of programs. In a paged computer, this characteristic is utilized in that a single memory-referencing instruction is designed to address *directly* only locations contained in the page in which the instruction is located and any location within *one* other page. Generally, this other page is the first page in the memory and is called the "base" page.

Most of the early minicomputer designs used exactly the method that we have just described for paging. An additional bit in the instruction word was reserved to indicate whether the page being addressed was the base page or the page containing the instruction. If it was necessary for the instruction to address a location that did not lie within one of these two pages, the location could be addressed indirectly via an intermediate location contained within one of these two pages.

Some of the newer computers employ an interesting variation on this technique, using what have come to be called "floating" pages as opposed to the older technique of "fixed" pages. With this technique, the page containing the instruction is not taken to have fixed boundaries—rather, the page is defined as being the contiguous set of memory words of which the current instruction is a middle word. Thus an instruction might address directly any word from its location minus $2^n - 1$ to its location plus 2^n, where n is the number of bits provided for its address field. Thus the "local" page "floats" along with the current instruction as the course of the program moves from one instruction to the next. Paging techniques are illustrated in Figure 4.10.

Now that we have discussed the addressing of data and operands, we turn our attention to the instruction set and see what some typical computer instructions do.

We noted three cases of instructions that addressed an operand of one kind or another. In one case, the operand was contained within a core location. In the second case, the operand was contained within a register in the central processing unit. In the third case, the operand location was an input/output device. The instruction sets of digital computers generally group instructions according to these classifications, and several others. We now consider a grouping that might be used for a typical modern digital computer applicable to computer-aided manufacturing problems.

1. *Memory referencing instructions.* This group of instructions always contains, at the very least, instructions for loading data from the core memory into registers and for storing data from these registers into the core memory. It may also contain instructions that increment or decrement the contents of memory locations. Occasionally, in 16-bit word machines, instructions are provided to handle halfwords (bytes). Some computers provide for the automatic comparison of the contents of a register and the contents of a core location, setting a detectable flip-flop if they are equal.

2. *Register referencing instructions.* These instructions operate only on the registers within the CPU. The group consists principally of instructions that shift the contents of the registers, increment or decrement the registers, complement them logically, clear them, or transfer data between the registers (if the architecture includes more than one programmable register).

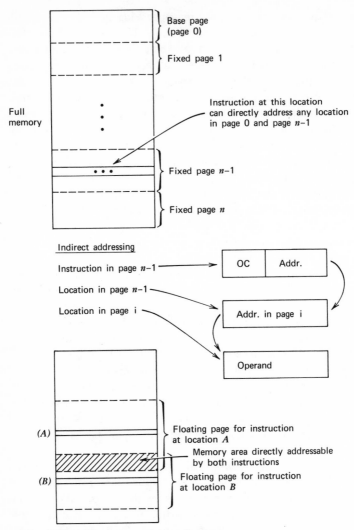

Base page
(page 0)

Fixed page 1

Full
memory

Instruction at this location
can directly address any location
in page 0 and page $n-1$

Fixed page $n-1$

Fixed page n

Indirect addressing

Instruction in page $n-1$

OC   Addr.

Location in page $n-1$

Location in page i

Addr. in page i

Operand

(A)

Floating page for instruction
at location A

Memory area directly addressable
by both instructions

(B)

Floating page for instruction
at location B

Figure 4.10. Paged memory organizations.

3. *Control instructions.* These instructions are used to control the flow and progression of the program. We noted earlier that the CPU retrieves instructions from the memory sequentially. In fact, a special register (called the "program counter" or the "location counter") is maintained by the CPU logic. This register contains the address of the next instruction to be retrieved from memory and is incremented as each instruction is executed. However, it is occasionally the case that the program must take

some alternate action based, perhaps, on the value of some item of data. For this purpose, instructions are provided that cause a "branch" in the instruction sequence, depending on the presence or absence of some condition. They operate by directly modifying the location counter. For example, the instruction might specify a transfer to a particular location within the memory if the content of the A accumulator is negative. In some of the more primitive computers, this branch is provided by causing the computer to skip the next instruction in the sequence through an extra increment of the location counter. An additional instruction is provided to transfer to a specified location unconditionally. Such an instruction, combined with a skip instruction, provides a full transfer capability in computers whose conditional transfer instructions do not provide address fields. (Of course, the unconditional transfer instruction must provide an address field—possibly paged.) We find that this kind of branching technique is frequently used in minicomputers that employ a form of "microprogramming." We discuss microprogramming later in this chapter.

4. *Input/output instructions.* This group of instructions is provided for communication between the computer and the peripheral devices and machines that it controls. Generally, this set includes instructions that cause the transfer of data from CPU registers to input/output devices or vice versa. Instructions are also provided that test the condition of the device—whether it is ready to transfer data, whether an error condition exists, and so forth. If the computer is equipped with a direct memory access channel, the counters and addresses utilized by this channel are initialized through transfers from the CPU registers. Finally, input/output control instructions are provided that command the device to initiate some action: transfer a block of data using the DMA, rewind a tape, and so on. The interrupt mechanisms of the computer are generally associated with the input/output system, and instructions that control interrupts are included in this group. These instructions allow and disallow ("enable," "disable") interrupts from devices, either as a group or individually.

It is a general principle that the more kinds of instructions available with the computer, the more powerful the computer. We may still find mildly amazing the amount of work that can be done by a very simple computer with few instructions—when it is properly and cleverly programmed. The reader may find it interesting to explore the programming possibilities of the following extraordinarily simple computer:

> The computer has only *one* instruction. The instruction has three address fields, call them A, B, and C. The instruction works in the following way: the contents of location B are subtracted from the contents of location A and stored back in location A. If the contents

of location A are now equal to zero, the next instruction is taken from location C. If not, the next instruction is taken from the location following the current instruction, that is, in the normal linear sequence.

The reader can gain a great deal of insight into the work that a programmer does by trying to code a few simple programs using this single instruction. For example, he might try to write a program that computes the sum of three or four numbers contained in the memory.

As our final topic in the category of classical architectures, let us discuss the register structure of the computer. Early minicomputers were provided with a single register in which arithmetic and logical operations could be carried out. The dimension of this register was equal to the width of the memory access port. There was also a single flip-flop register connected to this accumulator which was used to detect arithmetic overflows and as a temporary storage for a single bit which could be shifted out of the accumulator and into the single flip-flop register. Thus the single-bit register served as an extension of the accumulator.

Most small computers are now equipped with one or more registers in which arithmetic operations can be carried out. Generally at least two such registers are provided. Some of the more expensive minicomputers provide general working registers in groups of sixteen, imitating the register arrangements provided by some of the larger commercial and scientific general-purpose computers, such as the IBM S/360. If any of these registers can be used to modify the operand address specified in an instruction, then they are usually called "index registers." When a program must operate on an array of data stored in the core memory, the address of the first item in this array can be set into the address field of the accessing instruction and the contents of the index register can be used as a modifier, added (by CPU logic) to the address field to get the actual operand address. To obtain the next operand in the array, the program causes the index register to be incremented by one. The value contained in the index register can also be tested by the program to tell whether it has processed all of the items in the array.

In general, extra holding registers with arithmetic, shift, and indexing capabilities are very desirable attributes of minicomputer architecture. There must, of course, be at least one accumulator register. Most minicomputers on the market provide two registers with some means of shifting data from one into the other. A few computers provide either four or sixteen registers.

In considering a computer for an actual application, one should consider the complete economics of the problem. The tendency is for the more expensive computer to have more registers and be (generally) easier to program, so that the software costs are somewhat lower. However, if many copies of the system are to be made, the cost of the computer is the principal concern and a higher programming cost can be distributed over the cost of many systems. If only one example of the system is to be built, it may be less costly to spend a little more on a computer with a more powerful and easily programmed architecture; thereby playing software costs against costs of hardware.

Microprogrammed Computers

Microprogramming has been under consideration as a central processor design technique since the early days of computer development. It seems to have been first proposed by M. V. Wilkes in 1951. Briefly, and perhaps oversimply stated, a microprogrammed computer is one in which the instruction itself selects a *variable* sequence of extremely simple CPU logical operations for execution; this sequence is variable through some form of programming operation, as opposed to the fixed sequence selected in a hard-wired CPU design.

Some of the early minicomputers used a variant of microprogramming in several of the instructions with which the computer was equipped. For example, instruction "classes" for testing and altering the contents of registers were selected by one of the code values of the operation code field. The remainder of the bits in the instruction were given over to specifying whether or not an operation corresponding uniquely to that bit was to be performed. This is essentially the technique of elucidating on a basic class of instructions that we discussed before. For example, if bit 5 in the instruction were set, the computer would increment the contents of the accumulator. If bit 6 were set, the computer would complement the contents of the accumulator. If bit 7 were set, it would shift the accumulator one bit to the left, and so on. By coding the bits of this single instruction in the proper way, the programmer could cause a number of useful things to happen—all with one instruction execution. He might, for example, set up the instruction so that it would take the logical complement of the number in the accumulator and at the same time increment it thereby obtaining the negative of the number in the twos complement notation. In a similar way, instructions that tested the values of registers could be set up to test for a variety of conditions. For example, by setting micro-instruction bits so that a branch would occur if the contents of a register were not zero *and* not negative, the programmer could cause the computer to transfer if the contents of the register were strictly positive.

These versions of microprogramming were quite simple and for the most part only involved a small amount of direct gating logic, while adding quite a few useful instructions to the repertoire of the computer.

A more recent (and rather momentous) development has been the use of read-only memory (ROM) in microprogrammed computer design. The first major application of microprogramming for a computer that came to be in widely distributed use was in the IBM System/360. Since that time, much of the general approach used for the System/360 has been adapted to the design of smaller computers, as well as to that of other large commercial and scientific computers.

In most of these microprogrammed computers, the operation code no longer selects a particular hard-wired gating system output. Instead, it selects a small program (a microprogram) which is stored in a controlling ROM that is separate from the main core storage of the computer. The parameters used by this microprogram are a part of the instruction that selected it and the microprogram may also look elsewhere in the core memory for other parameters. The critical feature here is that the micro-program can interpret the bits of the instruction word (and any other applicable words or registers in the computer) in any way.

The instructions stored in the ROM are selected from a rather primitive set of fundamental hard-wired instructions out of which more complex operations can be fabricated. The instructions provided in the set might consist of single-bit shift commands, test and skip instructions, simple load and store commands in which the address of the operand is provided by the contents of a special register which the microprogram must set up in advance. By including the simple, single-bit, logical functions (AND, OR, complement), the microprogram may also carry out a binary addition operation on a bit-serial basis (albeit rather slowly), thereby eliminating the requirement for an adder circuit to be provided as part of the CPU hardware. The structure of an imaginary microprogrammed computer is illustrated in Figure 4.11.

Since a ROM is much faster than an ordinary core memory, several microinstructions can be executed in the time that would be required for the execution of an instruction in a computer of classical architecture. Thus the execution speed of a microprogrammed computer can be made comparable to that of an ordinary classically structured computer.

Perhaps the principal advantage of microprogramming is its flexibility. Many of the present microprogrammed computers utilize selectively wired memories in which U-shaped magnetic cores are employed. These memories can be easily and inexpensively wired by numerically controlled machines. The ROM is packaged so that it can be easily removed from the computer. Thus a user can design his own instruction set by writing small micropro-

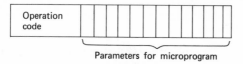

Parameters for microprogram

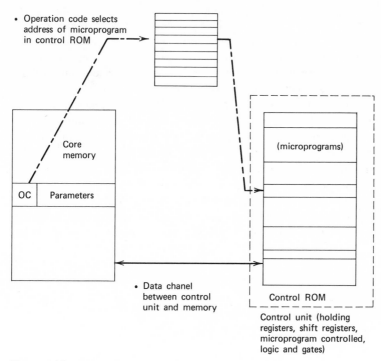

Figure 4.11. Microprogrammed computer structure.

grams (one for each of his own special instructions). A computer program can then be used to convert these microprograms into a punched or magnetic tape which controls the memory wiring machine. The newly wired ROM can be plugged into the computer and the user has a computer with an entirely new set of main instructions, specially tailored to his particular application. This was a particularly attractive feature of the System/360 when it was first introduced. The fact that the S/360 has a character-oriented memory and was microprogrammed meant that it could easily be set up to "emulate" older, character-oriented, business computers such as the IBM 1400 (in which many firms had a substantial software investment).

Decentralized Structures

Computers featuring decentralized structures are one of the most recent advances in computer architecture. With a computer employing a decentralized architecture, the concept of processing is significantly different from that used in conventional and microprogrammed computer structures. The processing idea (or "unit step," as it were) is to take an operand from some point, perform a predefined operation on it, and place it at some point. In more conventional structures, the process is broken down into an arrangement where each of these steps is a separate operation, perhaps requiring several instructions each. In a computer with a decentralized structure, all three steps are treated in one fell swoop. The variations in the instructions consist of specifying values for the two points and the predefined operation. While the operand is at a particular point, it may be acted on or processed in parallel with other operations.

To make this concept clearer, the reader should refer to Figure 4.12. We have an illustration of an imaginary decentralized computer architecture. A collection of devices is connected between two data transfer busses. Data, operands, and control signals are provided to the devices from the upper buss. Outputs of data and control signals are presented on the lower

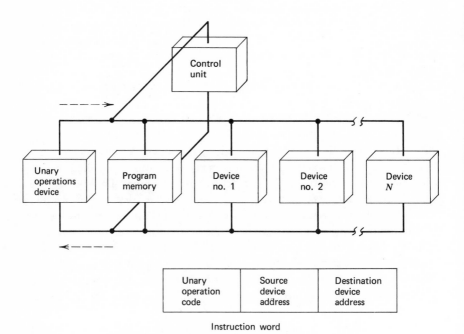

Instruction word

Figure 4.12. Decentralized computer structure.

buss. A collection of control logic is provided that retrieves instructions from a memory device and sets certain fixed instruction fields into control gating registers. The format of this instruction is shown in the figure. One field contains the address of one of the devices (or points, as we called them earlier). One word of data (or a group of control signals) is gated from this device by including its address in this field of the instruction. It can be termed the "source" device. The data word or group of control signals obtained from this device is routed to the device whose address appears in the second address field. This device can be termed the "destination" device. Once a device is given an operand or a control signal group, it performs some predefined function and, if there is a resulting output from this function, it sets up the output in an output register to await gating onto the lower buss by some subsequent instruction.

To obtain additional economy of operation, certain unary operations that might be performed on an operand can be provided by an additional device through which data always pass. This device can be controlled by a third field within the instruction. If the field is set up to enable the device, it might be used to complement the data word as it passes through, or to shift it to the right or left. Other useful functions such as overflow detection and extension bits, which were mentioned earlier, may also be provided.

One obvious advantage of the decentralized structure is its flexibility. This is an advantage that it shares with the microprogrammed architecture. If a special application-oriented function is needed, a device can be built to provide this function and it can be plugged into the buss structure of the decentralized computer. But the decentralized structure has an additional advantage that is largely economic. If the application to which the computer is being applied is quite simple and does not require extremely fast response times, only a few devices need to be included in the structure—devices that provide only some of the more fundamental operations and a few special ones. The program can then fabricate the control processes from these simple operations. If more speed and power are required, additional devices can be included (at additional cost, of course) that amplify the capability of the computer.

The principal application of these computers is expected to be largely in the control area, rather than as general scientific or commercial computing tools, since they are more closely related to programmed controllers than to computers, in a functional sense.

Input/Output Structures

If the results of the work that a computer does (its output) cannot be communicated to the world outside the computer, then the work is of no value. The outputs must be either reported to the user in human-readable form or delivered to an external device or machine where they specify some control function or are retained for later use. Similarly, the computer must be able to react to stimuli that represent events and conditions in the machines with which it communicates. The purpose of the input/output structure of the computer is the satisfaction of these requirements.

There are many approaches to the computer input/output problem and a great deal of often confusing terminology is associated with the hardware and techniques that are employed. Our purpose is to give the reader some information on the general, and typical, solutions that are employed in this area. When he finds himself dealing with an actual computer vendor or an actual or proposed computer application, he may find it necessary to learn the meanings of the terminology as it is used in that context. He will usually find that the principles remain much the same—only the names have been changed to promote the interests of the vendors and designers.

We begin by discussing four topics that fall under the category of input/output structures:

1. Programmed input/output
2. Interrupts
3. Direct memory access
4. Input/output processors and channel controllers

We then discuss the technology of some of the more important computer peripheral devices.

Programmed Input/Output

Programmed input/output is the most fundamental form of data transfer, and every computer is equipped with some capability for data transfers using this technique. The technique is characterized by the direct attention of the central processing unit to each step of the data transfer. In a typical output transfer, for example, instructions sense the readiness of the peripheral device for the transfer, place the item of data to be transferred in one of the CPU registers, and then command the transfer. In some cases, the computer must hold the data in the register until the peripheral device signals the computer that it has accepted the data. Other minor variations exist. For example, the data item may be in one of the core memory words when the transfer instruction is executed. The CPU logic provides for retrieving the word from the core memory location and placing it in a holding register while waiting for the peripheral device to accept the data. This variation is largely a convenience for the programmer. The programmed transfer characteristic is still present. That is, the data transfer is synchronized with the execution of the input/output instruction by the central processing unit. For input transfers, the instructions sense the readiness of the device, command the transfer, wait until the data has been gated into the CPU register, and then store the data in the core memory or proceed to process it.

Figure 4.13 illustrates the components and steps in a typical programmed data transfer in which a word is to be sent from the computer to a peripheral device. The principal lines of communication and data transfer are contained in an input/output buss which leads to all of the peripheral

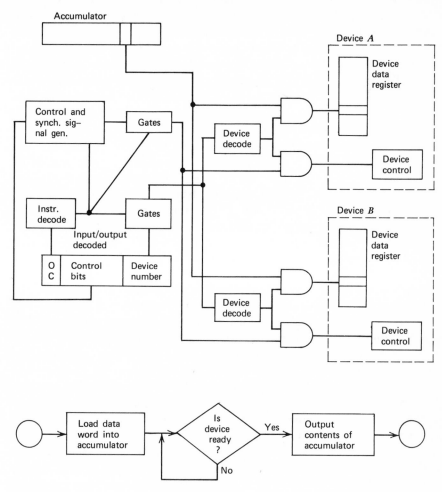

Figure 4.13. Typical programmed input/output.

devices. This buss may be entirely within the CPU rack or it may be a cable, daisy-chained to each device in sequence. This buss contains data transfer lines and lines for control and synchronization signals.

The form of a typical programmed input/output instruction is also shown in Figure 4.13. Generally, such an instruction has three principal fields. The first field is the operation code which identifies the instruction to the CPU as an input/output control or data transfer instruction. A second field within the instruction bears the code number of the device to which the command is addressed. The third field is used to specify variations on the basic input/output instruction. For example, test the status of the device, command a function (start, stop, rewind, and so on), transfer an item of data to the computer, or accept an item of data from the computer. In the simplest (and most typical) programmed input/output designs, the contents of the control bits and device code field are placed on the input/output buss and it is up to the peripheral device to decode them and respond properly. The CPU usually provides a synchronization pulse that the selected peripheral must use to strobe data on and off the buss. In some models, the control bits in the instruction enable a sequence of control pulses on separate lines, each line being associated with one of the control bits. The programmer may set these bits in the instruction so that they generate a sequence of control pulses that command the peripheral device to progress to a particular state, or to carry out some sequence of logical functions that results in the transfer of data. For example, one bit might be used to clear a holding register in the peripheral device. The second bit might cause a pulse used to gate data from the buss into the holding register. A third bit might cause a pulse that initiates the use of the data within the peripheral device; for example, punching the bit pattern into paper tape.

There are a number of variations on the scheme that we have just described; each having its own merits in terms of programming ease, simplicity of device interfacing, reliability and speed of data transfer, and so on. In computers with multiregister CPU structures, a field may be provided within the instruction to specify the register into which the data are to be transferred. The synchronization scheme between the computer and the peripheral device may consist only of a pair of flip-flops: one to be set by the computer when it wishes to send something to the peripheral, and the other to be set by the peripheral when it wishes to send something to the computer.

The potential user of a computer, especially for an application involving the control of machinery, should carefully study the requirements of interfacing his machinery to the various computers that he may be considering. With some of the computers now on the market, much of the decoding and control logic for programmed input/output is provided within the central processor package and within the basic CPU cost. With other computers, much of this logic must be provided by the user or is available separately from the computer vendor, but at additional cost. This is another example of a situation where the application and its total cost (both in hardware and engineering labor) must be carefully studied before selecting a final course of action.

Interrupts

As we mentioned earlier, an interrupt capability is generally provided with the computer so that a peripheral device can signal the CPU when some external event has occurred or when a certain condition has been attained. The principal advantage of this capability is that it allows the CPU to carry out other processing tasks while the peripheral devices are working and still be able to respond quickly to external events. The interrupt feature is generally used to signal the CPU that a peripheral device has completed the operation that it last commanded and is now ready for the next operation. For example, a disk unit might be ordered to transfer a large block of data into the core memory. The program would set up the transfer using the programmed input/output techniques, providing the disk controller with the count of words to be transferred and the address within the core memory where they are to be stored. The disk controller would then transfer the complete block using the DMA. While this transfer was taking place, the CPU could be carrying out other processing. For example, it could be calculating the data that are to go into the next block. When the disk finished the transfer, it would interrupt the CPU. This interrupt would signal the CPU that the transfer was complete. The CPU could then initiate the next transfer.

There are several things that a properly designed interrupt system must be able to accomplish. First, it should notify the computer promptly and in such a way that the computer will not fail to notice that the event has occurred. Second, there must be some means for the computer programs to determine exactly which peripheral device is calling for attention. Third, there must be some reasonable means of resolving conflicts if more than one peripheral device should ask for attention at the same time. Finally, the computer should have the ability to turn the interrupt system on and off at will—this is necessary so that the programs will not become confused by interrupts that occur at closely spaced points in time.

To illustrate the way in which a typical interrupt system might work, we will proceed through a sequence of steps that might occur in an actual situation and discuss some of the variations that are offered with different computer systems. First, let us suppose that there are at least two peripheral devices attached to the computer and that they are carrying out some operation that the CPU has commanded. Suppose further that both operations are completed at the same time. In the meantime, the programs in the computer have been running, carrying out some calculations or dealing with other peripheral devices.

As the two operations are completed, both peripheral devices note this and wish to raise signals to the CPU to cause an interrupt. The first thing that must be done is to determine which one goes first. This is the priority problem and there are several ways of resolving it. In the simpler minicomputer systems, it is resolved as illustrated in Figure 4.14. Each device is required to remember that it wants to generate an interrupt and the

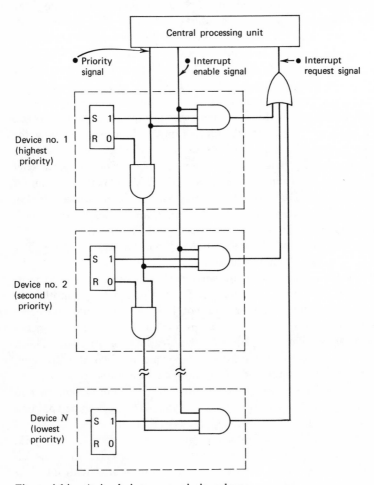

Figure 4.14. A simple interrupt priority scheme.

mechanism by which this is accomplished is a flip-flop that is part of the device itself. The CPU only allows interrupts to occur at times that are convenient to the sequencing of its logic (e.g., between instruction executions), and it synchronizes peripheral devices to this time by providing an interrupt enable signal on one of the input/output buss lines. It also provides an interrupt priority signal on one of the buss lines. Note from Figure 4.14 that the priority signal is routed from one device to the next. The device that gets the priority signal first is the device with the highest priority. If this device wishes to cause an interrupt, it can prevent the signal from being routed to the next device—one of lower priority. Thus the higher priority device causes the first interrupt. When its interrupt has been acknowledged by the CPU, it can then allow the priority signal to pass on to devices of lower priority so that their interrupt requests can be handled by the computer. With some of the more expensive minicomputers, this priority determination logic is provided within the CPU rack itself. With others, it must be built into the interface logic of the peripheral device or controlled machine by the user.

Once a device has received the interrupt enable signal and is requesting an interrupt, it can send an interrupt request signal to the CPU. This signal will cause the CPU to carry out a chain of events that alter the course of the program operating in the computer. First, the CPU will cause the program to execute an instruction stored in a predefined location within the computer memory. At the same time, further interrupts will be temporarily disabled. The CPU has a register that is generally called the "program counter" or the "location counter." At the time that the CPU permits an interrupt, this register contains the core memory address of the next instruction to be executed. In the normal course of events, this register is incremented by one as each instruction is executed. (The conditional and unconditional transfer instructions that we discussed earlier cause this register to be loaded with the address of the instruction to which the program wishes to transfer.) In the case of an interrupt, the CPU causes the instruction stored at the predefined location to be executed in the place of the instruction that would normally be executed next. The instruction that is stored at this location is an instruction which causes the computer to transfer to an address in the core memory and at the same time store the contents of the program counter. Generally, although not in all computers, this instruction stores the program counter at the address specified in the instruction and transfers control to the next location following this address. The result of this sequence is to cause the computer program to enter a special routine intended to recognize and respond to interrupts. When this routine has finished the processing required by the interrupt event, it can return to the main part ·of the program. It does

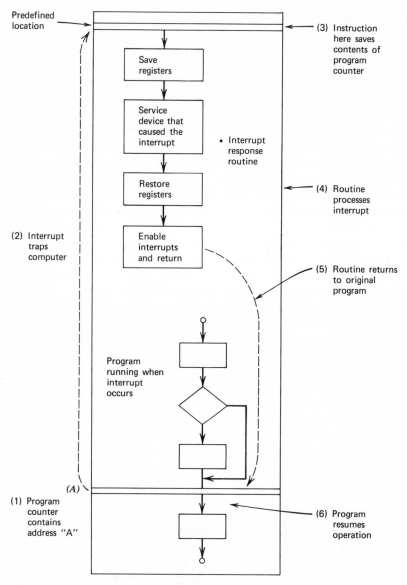

Figure 4.15. Interrupt response procedure.

this by transferring control to the location that was stored by the instruction in the predefined location when the interrupt occurred. This procedure is illustrated in Figure 4.15.

This interrupt has come at a time that is essentially unexpected by the program, and it is generally the case that the program will be carrying out some task not directly related to the interrupt. Thus the routine that is entered as a result of the interrupt is going to have to, more or less, put things back the way they were before the interrupt. That is, it is going to have to save all of the active, programmable, CPU registers and reload them with their original contents when it returns control of the computer to the interrupted program. Recall that the CPU disabled further interrupts when it began its response to the present interrupt. Thus no other interrupt can occur until one of the computer programs uses an interrupt enable instruction to allow a subsequent interrupt. The interrupt response routine uses the time during which interrupts are not enabled to store away the live registers (save the "machine conditions," in programming parlance). It may then enable further interrupts, allowing itself to be interrupted if another device should require attention.

We have said that the interrupt altered the course of the program by executing the instruction stored in a predefined location in the core memory instead of the next instruction in the normal sequence. In programming terms, this is called "trapping" the program. Depending on the design of the computer, there may be only one such predefined location or there may be many such locations, each one associated with one or more peripheral devices.

One of the principal problems with processing an interrupt is determining the cause of the interrupt. In the simplest computers, the program is trapped to only a single location, no matter what the cause of the interrupt. The program must then "interrogate" each peripheral device that is capable of causing an interrupt. This interrogation is done using one of the input/output instructions. During this interrogation, other interrupts must be disabled to keep the program from losing track of the original interrupt if another should come along. In some of the more sophisticated designs, a location is assigned to each device and the identification is derived from the location to which the computer program is trapped. In others, the device may be required to provide an identifying code when it requests the interrupt. This number is held in either the device or a CPU register until the program reads it and uses the number to identify the device, In some computers, the device actually supplies the core memory address to which the computer is to be trapped.

When an interrupt response routine has completed its processing, it returns to the main program. If the problem permits it, the interrupt response routine may enable further interrupts after it has saved the machine conditions. In our example, the second device waiting to cause an interrupt would then be serviced and a trap would occur within the first interrupt response program, immediately after it had enabled interrupts. This might trap the computer to a different location at which an interrupt response routine for the second device would be started. When this routine finished its processing, it would then naturally return to the interrupt response routine that was handling the first device.

In general, the more stringent the response time requirements, the more sophisticated the interrupt scheme should be—especially if there are several devices that require such response. There are many possible variations and almost every computer has its own approach to the problem. Some, for example, have made the problem of saving machine conditions much easier by providing a parallel set of programmable CPU registers. Thus up to one level of interrupts may save machine conditions by employing a single special instruction that switches the CPU to work with the second set of registers, reserving them, as it were, for use only by the interrupt response program. The interrupt structure is one of the most critical parts of a computer intended for machine control applications and it should be carefully examined before selecting the computer.

Direct Memory Access

We discussed direct memory access channels early in this chapter when we were surveying the general nature of a classical digital computer structure. A direct memory access channel (DMA in general usage in the computing community) is a data channel for peripheral devices that provides them with direct access to the core memory of the computer, so that data can be transferred without the intervention of computer programs except for the initialization of the transfer process.

The general core memory device provided with a digital computer requires two functional registers for its operation: one to contain the address of the location in the memory to which reference is being made, and the second to act as a buffer for the data whether it be incoming or outgoing. In addition to these two registers, control signals are provided that bear the instructions to the memory (read or write). In the simplest DMA channel designs, connector terminals to these two registers and to the control signals are provided to the external device so that a peripheral device can share control of the memory operations with the CPU. In order to resolve conflicts of access between the CPU and the peripheral device connected to the DMA channel, a "lockout" control signal is provided. When the peripheral device wishes to gain access to the core memory, it raises this lockout signal and its presence prevents the CPU from gaining access to the core memory until the peripheral device performs its transfer and releases the memory. This is the arrangement that was illustrated in Figure 4.1.

By its general physical nature, a core memory cannot permit simultaneous access to its contents if the addressed items of data are not stored in the same memory word. Thus cooperation is necessary between devices seeking to store or retrieve data from the same core memory. In some digital computers (generally the larger and more expensive models), the basic core memory is divided into several essentially independent sections. Each one is actually a separate core memory with independent address and buffer registers and control signals. With such an arrangement (called a multiport memory), it is possible for two different devices to gain simultaneous access to the core memory so long as they are referring to separate sections. Conflict resolution logic operates in a somewhat different manner than in the single-port case. Independent priority is not given unilaterally to one of the devices as in the case of the simple design discussed above and illustrated in Figure 4.1. Instead, the conflict resolution logic monitors the addresses provided by the two (or more) devices and looks to see if they represent words in the same section of the core memory. If they are not in the same section, the logic permits each access to proceed independently. If they are in the same section, then the conflict resolution logic holds off on one of the devices, allows the other to complete its access, and then permits the remaining access to occur.

Generally, access conflicts are resolved in the favor of the peripheral device rather than the central processing unit. This is because peripheral devices attached to a DMA channel usually transfer highly synchronous or highly transient data. For example, a magnetic tape or a disk memory must get the next data word in time to write it on the portion of the recording medium that is currently under or very near the recording head. The data are written "on the fly" and the recording medium is generally moving at a very fast rate. If data going to this device are held up too long, the section of the recording medium on which it is to be written will have passed under the head and a gap will be left, or the synchronization of the device will be lost. The CPU, on the other hand, is executing program instructions stored in the core memory. In general, the individual instructions of a program need not be precisely synchronized; they will not be lost if their retrieval is momentarily delayed nor will the CPU lose the opportunity of storing an item of data in the core memory.

In addition to the access channel that we have been discussing, the general process of direct memory access data transfer requires two items of information: the location in the core memory at which the data are to be retrieved or stored, and the number of words that are to be transferred.

Thus logic must be provided to count the number of words transferred and to increment the core memory address as each word is transferred. A means must be provided whereby this logic is given the starting address and the word count. Usually, the starting address and the word count are provided via the programmed input/output transfer technique. Once the program has transferred these items of data to the DMA logic, it commands the DMA to initiate the automatic transfer of the block of data. When the DMA has transferred the requested number of words, it may notify the program by generating an interrupt.

Depending on the price and design of the computer, the DMA logic may provide either no support to this function or a great deal of support. Some computers advertise DMA channels as part of their capability when, in fact, all that is provided are the access terminals and connectors for the data, control, and conflict resolution signals. All of the address calculation, word counting, and interrupt request logic must be provided as part of the logic associated with the peripheral device to which the DMA is connected. With other computers, the counting and address logic is provided either as an option or as an integral part of the CPU logic and packaging. Some DMA channels provide for special processing of the data as it is being transferred. For example, some peripheral devices record data in the form of bytes (8-bit characters). In a 16-bit computer, the data are generally stored two bytes to the word and the DMA channel takes care of unpacking or packing the data with respect to this form of storage. Some DMA channels are capable of transferring data from more than one peripheral device. It is switched from one device to another by instructions in the computer program.

Input/Output Processors and Channel Controllers

The next level of sophistication beyond the DMA channels that we have been discussing is represented by the input/output processor and the channel controller. These devices are small computers in their own right and perform the specialized function of handling all input/output for the computer system, working in parallel with the central processing unit. Among minicomputers they are very rare, but they are *de rigeur* for large-scale scientific and commercial data processing systems.

These devices are provided with a direct memory access to the core memory of the computer and are initialized in much the same way as the DMA channel, with certain minor exceptions. In the simplest designs, programmed input/output transfers take care of this initialization. However, in the large-scale systems, the decoding of an input/output instruction by the CPU automatically brings the input/output processor into play and the details of the instruction are provided by the CPU for further decoding and execution. Then, while the CPL continues to execute the other instructions of the program, the input/output processor starts to carry out the input/output instruction.

The input/output processor uses its DMA channel for two purposes. The first is the transfer of data to and from peripheral devices—the DMA utilization that we have already described. Its second use is the retrieval of special instructions for execution within the input/output processor. The execution of an input/output instruction within the CPU generally provides the input/output processor with an address in the core memory at which an entire special computer program is stored. This program contains instructions for the transfer of data and the control of peripheral devices, and these instructions are structured in a special format oriented for interpretation and execution by the input/output processor rather than the CPU. The processor reads these instructions from the core memory in much the same fashion as the CPU obtains its instructions from the core memory. It carries out the processing implied by these instructions in a sequential fashion, until it comes to an "I/O halt" instruction. It then waits for the next input/output initiation instruction to be executed by the CPU to start the process over again, perhaps this time with a different input/output program for another device. With some input/output processors, branch instructions and subroutine entry instructions are provided so that very flexible input/output programs can be written. The general operation of such a processor is illustrated in Figure 4.16.

The advent of inexpensive ROM minicomputers and controllers may make input/output processors, such as we have just described, available as reasonable options for minicomputers. These ROM processors would be set up to work from the DMA channel of the host minicomputers and would be programmed to interpret programs stored in the core memory of the host computer. Present market costs of ROM controllers capable of this function are about $3000 and they present intriguing possibilities to many minicomputer manufacturers.

Peripheral Devices

Computer vendors generally provide a line of optional peripheral devices for use with their computers. Usually, these devices are manufactured by companies other than the vendor of the computer. With most potential computer applications in the direct control of machines, the principal amount of attention is paid to the capabilities of the central processing unit and the core memory and the interface between the CPU and the machines that it is to control. The principal application of standard computer peripheral devices is for data logging, loading of programs, preparation of programs, and the storage of data bases. Although quite important,

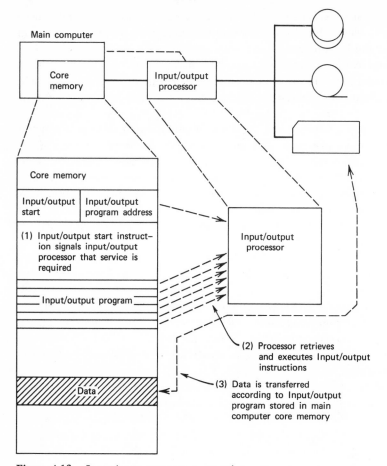

Figure 4.16. Input/output processor operation.

they are unfortunately frequently relegated to positions of secondary consideration.

We now treat the technology of some of these peripheral devices so that the reader has some appreciation of the terminology, capabilities, and restrictions of these devices.

With most minicomputers, the basic input/output device is the Teletypewriter Model ASR 33 (ASR stands for automatic send-receive). This device includes a keyboard with a full alphabetic and numeric character set (plus certain special control characters), a printer, a punched-tape reader, and a paper-tape perforator. The basic data rate for this device is ten characters per second, and the method of data transmission to and from the Teletypewriter is bit serial at a baud rate of 110 bits per second. There are actually eleven bits in the transmitted character, the character being marked at the beginning and end with start/stop-bit patterns so that data transmission can be asynchronous. The Teletypewriter is frequently used for the preparation of computer programs, punching them onto paper tape for permanent recording and loading them through the punched-tape reader.

Punched paper tape is the principal input/output medium for small computer systems, largely as a result of the use of the Teletypewriter, which is relatively inexpensive. Programs are prepared in human-oriented symbology and converted to computer-readable and executable instructions using other computer programs. These programs are called "assemblers" and "compilers" and we discuss them shortly. It should be noted now, however, that the output of these special utility programs represent the actual operating application programs and much of the documentation for these programs. As a result, the computer system and its operators frequently must handle a great deal of punched tape (generally more than they may have imagined when they first began, especially if they are inexperienced). Reading and punching of paper tape at ten characters per second is quite slow and completely unacceptable for applications where the programs exceed more than a very modest size. Thus higher-speed punched-tape handling equipment is generally provided. This includes readers with rates of computer input of up to 300 characters per second and paper-tape punches with typical rates of from 60 to 100 characters per second. Since tapes are read many more times than they are punched, high-speed readers are more prevalent than high-speed punches.

For minicomputers in control applications, other principal peripheral devices are those that represent magnetic recording media, usually magnetic tapes and disks and drums. The major varieties are listed below with brief descriptions of their operation:

1. *Conventional magnetic tapes.* These operate by altering the magnetization of oxide-coated mylar tape wound in reels of varying capacity. The data are recorded in a uniform and synchronous manner, and the tape is continually moving during read/write operations. Data are recorded in character form, one character being written across the short dimension of the tape. Characters of either 7- or 9-bit lengths are recorded, depending on the design of the tape unit, with one bit being reserved for lateral parity on the recorded character. Blocks of data are written in records, the tape being brought up to speed by the controller before it accepts data from the computer. When a complete block has been written, the tape is brought to a stop after a longitudinal parity character is written

and an end-of-record mark is placed on the tape. Usually, other special marks are provided to mark the end of a "file"—a collection of related records. Typical data transfer rates vary from 12,000 to 24,000 characters per second or more. Character packing densities vary from about 200 to 1600 characters per inch of tape.

2. *Incremental magnetic tapes.* These devices use approximately the same recording technology as a conventional magnetic tape. However, each character is recorded as it is received by the tape drive with the tape being moved forward in one complete start/stop cycle for each character, hence the name "incremental." These devices do not automatically provide the parity checking features of conventional tape units and do not require the same degree of logical complexity and timing precision. For these reasons, they are somewhat less expensive than conventional magnetic tape drives. On the other hand, their recording capacity and rate of data transfer are much smaller. They are generally used for applications in which a moderately large amount of data must be recorded—not enough to justify a conventional tape drive, but too much for punched tape to be an acceptable recording medium.

3. *Cartridge and cassette recorders.* These devices are a relatively new addition to the field of computer peripheral devices. They are actually small magnetic tape drives in which cartridges and cassettes are used to hold the tape, instead of reels. They are similar to units used in home high-fidelity music systems. Their capacity is quite limited, generally to a few hundred thousand characters per cartridge or cassette, and their transfer rates are usually about equal to those of high-speed punched-tape readers (about 300 characters per second). The cost of one of these units is approximately the cost of a high-speed punched-tape reader, generally a few thousand dollars. They are principally intended to serve as replacements for these devices.

4. *Disks and drums.* The recording surface for these devices is either a rotating platter (in the case of the disk) or a rotating cylinder (in the case of the drum). The rotating element is also coated with an oxide material susceptible to magnetic recording of digital data. The tape units that we discussed are sequential devices, with the data recorded from one end of the tape to the other. With disks and drums, data are recorded in tracks that encircle the recording surface in the direction of rotation. Read/write heads are positioned over these tracks and the data are read or written as the track moves under the head. Clock tracks are provided that permit coordination of the writing operation with particular physical locations on the recording surface. Thus these devices are highly synchronous. Their rotation rates generally bring every portion of the recording surface within one track under the read/write head at least once every

33 to 67 milliseconds. Some disk units stack the recording platters in groups of six to ten platters, all assembled as a single removable unit. The read/write heads are mounted on a moving arm that is withdrawn when the stacks, called "disk packs," are removed or mounted on the drive unit. The arm holding the read/write heads can be moved from one track to another under the control of the computer or the input/output processor when this kind of design is used. With single platter disks and with drums, a read/write head is generally provided for each track on the recording surface. Disks and drums are characterized by the name "random access device" and "direct access device" in the computing trade terminology. They are available in a wide variety of configurations, speeds, and capacities that vary from a few thousand words up to several million words.

5. *Card readers and punches.* These devices use the ordinary ubiquitous IBM card for their recording medium, either punching the holes in the cards or reading them. They are usually associated with data processing applications and are rare in computer systems used for control applications. Generally, they are used in place of punched-tape equipment for the preparation of programs and their handling. Occasionally, they are used for data entry, for example, in warehouse applications where stock orders are prepared in sales or purchasing departments on ordinary keypunching equipment and forwarded to the warehouse computer.

6. *Other graphic and printing devices.* Another type of peripheral device frequently found with a minicomputer installation or a computer control system is the line printer. These devices are used in program preparation and in the logging of large amounts of data. They are high-speed printing devices, generally equipped with a buffer that holds a line of characters. This buffer is loaded at high speed directly from the computer and printed on the paper in a single line when the buffer has been filled.

Less frequently, the computer may be equipped with a plotting device or a CRT device. We discussed the use of plotters in Chapter 3 when we discussed the computerized drawing of wiring diagrams, printed circuit board layouts, and schematics. Cathode ray tube graphic devices are usually used in computer-aided manufacturing applications for the temporary display of similar graphic material so that the engineer can check it and correct it, if need be. Cathode ray tube devices generally fall into two principal categories: those that can "draw" lines and those that can only display characters (as a rule, through the use of a matrix of illuminated dots on the tube face). The former category is of more interest in computer-aided manufacturing. The latter are most often used in business-oriented applications where readable data is to be presented temporarily but no hard copy record of the data is required.

Advanced Architectures

The architectural features that we now discuss are not generally available in computer configurations intended for control applications. They represent approaches to advanced computer design that have actually been implemented only in a few installations, mostly either extremely costly data processing facilities devoted to scientific information processing or in experimental facilities in universities or manufacturer's R&D departments.

Most computer systems consist of a single central processor, a core memory, and various peripheral devices and interfaces to controlled machinery. In some advanced and theoretical computer architectures, the central processor may, in fact, consist of several separate processors acting in concert. In the conventional computer system, the central processor retrieves each word from the core memory as it is required by the process. In some advanced configurations, there may actually be a buffer store standing between the central processor and the main core memory; the purpose of the buffer is to speed up the execution of the programs running in the computer. The contents of the ordinary core memory are addressed by providing the location within the memory that contains the data item in which we are interested. Another form of memory provides for the retrieval of an item of data by providing the name of the data rather than its location in the memory.

We treat two categories within this subject of advanced architectures: memory organizations and multiprocessor systems. As to their relevance to the subject of computer-aided manufacturing, there are two points of contact. First, these advanced systems will be found in increasing frequency in the large data processing installations and service centers of the future. People working in the area of design automation and the development of manufacturing data bases will find themselves running their programs on these advanced machines. Many of the present day computational problems in design automation may be relieved through the application of these computers. As the second point of contact, some of the architectural features of these advanced machines will soon find their way into the architectures of the smaller computer systems used in the direct control of manufacturing equipment.

Memory Organizations

The classical architecture, as well as the microprogrammed and decentralized machines, all feature a single core memory from which the CPU retrieves single words as it proceeds to execute the instructions that make up the program. Core memories exhibit certain physical limitations as to highest attainable speeds. In general, much greater speed could be attained from integrated circuit memories. But integrated circuit memories are relatively expensive when compared to core. The question then is whether some architectural arrangement or mix of the two kinds of memories could perhaps improve the overall processing speed of a computer. An attempt to answer this question is provided by equipping the computer with high-speed integrated circuit buffer memories of limited capacity and using the conventional core storage techniques to make up large memories from which these buffer memories are fed. Figure 4.17 illustrates the arrangement in

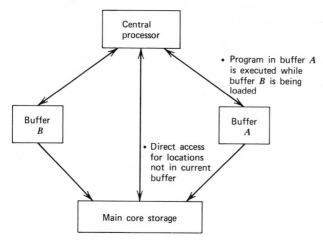

Figure 4.17. Buffered memory architecture.

such a computer design. Here, two separate buffer memories are provided. Programs are executed only from one of the two buffer memories. Each buffer memory can be thought of in much the same way as the pages of conventional memory which we discussed earlier.

Just as the page concept in conventional memories has an impact on the design of computer programs (because of the restrictions imposed with regard to direct access to data in other pages of memory), so the buffer memory concept affects the design of programs intended to operate on computers with this kind of architecture. For example, access to data words not contained in the buffer memory is clearly less efficient than access to words that are presently in the buffer. Thus these memories are more efficient when applied to problems in which only a few items of data are required. It is also apparent that programs which can be wholly contained within one buffer memory module will run faster than those which exceed the size of one module.

The general operating procedure for a buffer memory architecture is as follows:

1. One buffer, buffer A, is filled from the main core storage and the program contained therein is started.

2. While the program in buffer A is running, the second buffer, buffer B, is loaded in parallel.

3. When the program in buffer A has finished its processing, then the program in buffer B is started.

4. While the program in buffer B is running, buffer A can be filled from the main core storage with a third program.

Naturally, the logical functions provided in the CPU are much more complex than those provided in a classical configuration. For example, the CPU must be able to recognize a reference to a memory location which has not been transferred into the buffer storage and to handle this reference in some way—perhaps by operating in the normal manner for that reference; that is, going directly into the main core storage for that one word.

The theory of writing programs for computers with this kind of memory arrangement is the subject of much study by computer science theoreticians. The improvements in efficiency for certain classes of computational problems are examined with respect to the various ways that the CPU might be designed to handle the filling and writing back into main core memory of the various buffers. The effect of additional buffers is also studied.

We have noted that these buffer memory architectures are efficient for computational problems in which all of the required data can be carried along when the buffer is loaded with the program. What of problems that require access to large amounts of data that may not be organized in the most optimal way? For example, information retrieval systems in which the programs must look up information would be more efficient if they could directly ask the hardware to provide a collection of data identified by name. Rather than searching through a large file for the records associated with all persons with a certain characteristic, if the memory could be addressed by the characteristic, the record could be found more promptly.

As an example, consider the problem of finding the record for a given part number in an inventory file. Each part number record contains such data as on-hand quantity, warehouse location, cost, economic order quantity, reorder point, and so on. Suppose a warehouse manager wished to inquire about the status of a particular part number in his inventory. He would command the computer to find and print the information for this part number. The inventory file would be referenced by a directory that contained all of the part numbers in the inventory and pointers (addresses in the main inventory file). If the directory was not ordered in any particular way, the computer would have to successively examine each directory entry until it found the entry for the part number named in the inquiry. Only then could it obtain the record containing the actual information. This may imply a large amount of time spent searching the directory. If the directory were ordered, for example, in alphabetical or numerical order, then more efficient search procedures could be used, but a good deal of time would still be consumed. If, through the use of special logic, the memory could be provided with the part number, it could then output the pointer immediately. That is, the address in the memory would not be an actual location number but would instead be the name of the item of information desired. A memory with this kind of addressing scheme is called an "associative" memory.

There are various ways of designing the hardware that forms an associative memory, and we do not propose to explore all of them. The basic concept, however, is to store the name of the data items as well as the data item and to use high-speed electronic techniques to make an exhaustive parallel search of the memory, looking for a name entry that corresponds to the address provided by the user of the memory. Associative memories are still highly experimental, but it is expected that the large-scale integration techniques may make them readily available in the future.

Multi- and Parallel-Processor Systems

In the conventional computer architecture, we have only a single central processing unit carrying out work at one time. When an input/output processor is included in the configuration, it is generally for the convenience of the data transfer problem and is not formally considered to be an element of a multiprocessor system. For most computational processes and control problems, a single fast processor can handle all of the work, but it must carry out this work in a sequential way, one instruction at a time. In the hope of attaining faster processing and response speeds, the concept of the multiprocessor architecture was developed. Some computer vendors presently offer systems in which one extra processor is included. In a typical application, one of the processors would act as an executive and control device, while the second might carry out miscellaneous processing tasks in parallel with the operation of the first processor. This kind of configuration does not represent any great advance in CPU design since it is generally one in which we have two computers with some form of high-speed data channel connecting them. They can operate from the same core memory, sharing it in the way that a conventional processor and a DMA would share it. Usually, the advantages of such an architecture are dependent on the problem being solved.

Some of the modern computational problems (such as graphics

processing, time sharing, and signal processing) require the handling of large data matrices. Many of the operations performed on these matrices, such as matrix multiplication, can be broken down into a number of parallel operations, each operation working on a subset of the matrix, with the results of that operation not being required for any of the other steps involved in the procedure. There are many mathematical processes that exhibit these characteristics, and the improvement of their efficiency is one of the objectives of parallel-processor design.

In a parallel-processor design, we actually have a large number of primitive CPUs carrying out operations in parallel. Some of the experimental parallel processors have as many as 256 or 512 adding circuits operating in parallel. The basic instructions to the controlling processor provide for the expression of matrix and array operations—an extension over the classical computer instruction which manipulates only one item of data at a time.

Again, as you might expect, using such a computer system has important ramifications in the design of programs and computational procedures for it. A great deal of research is still being done in these areas and it will probably be years before these computers are in widespread use. With respect to the field of computer-aided manufacturing, their principal application will probably be in the area of graphics processing for the design of interconnection systems and in the area of logical simulation for the testing of electronic system designs.

COMPUTER LANGUAGES AND SOFTWARE SYSTEMS

We know that a computer is a machine that carries out certain procedures and programs according to instructions fed to it from some source, generally a memory device associated with the processing unit. Without these programs and procedures the computer has no work to do, so we must consider them an essential part of a computer system. In fact, they are much more than that. Good programs and other supporting software serve to extend the capabilities of the computer system greatly. The separation of software and hardware considerations cannot be permitted in either the evaluation of a computer system or the design of the large system in which it plays a controlling role. In this section we discuss some of the aspects of computer software.

Programs are procedures. They are a collection of instructions directed toward some kind of processor. The processor carries out the operations implied by each instruction in a sequential manner. The instructions are expressed in a language that must have a syntax and a grammar. Some languages make it easy to describe certain kinds of procedures. In other languages, a particular procedure may be quite difficult to express. In general, there are several problems that must be addressed when considering the expression of a procedure. One of them is the question of ambiguity. This is an especially important consideration in the field of computer languages, since the computer hardware interprets each instruction given it in a sometimes distressingly literal fashion. Another question is the completeness of the language. The programmer must be able to express his procedure in the language, so it must therefore provide the proper words for making up the operations. Still another question is the conciseness of the language. How many words and sentences are required to express the procedure? Generally, the fewer the better for at least two reasons: (1) so that the task of writing out the procedure represents less work, and (2) because resources for the retention of the program are an economic variable in the computer system (memory costs money).

It is presently the case in the field of computer languages that no single ultimately desirable computer language exists. Work is being done to make languages as general as possible so that programs written in them can be used on almost every model of computer and so that every possible procedure can be expressed in this language in a way that is economically and efficiently reasonable. There are several languages that are widely accepted in general use in the fields of scientific and business computing, and some variants of the scientific languages are employed from the development of large-scale computer control systems. However, for the most part, there are almost as many computer languages as there are major human languages.

Each computer language belongs to one of three categories:

1. Machine languages (also called "object code")
2. Assembly languages
3. Procedure-oriented, problem-oriented, and algorithmic languages

Languages falling in the first two categories are almost 100 percent "machine dependent." That is, programs expressed in these languages operate on only a single model or line of computers. Procedures expressed in these languages are formulated by writing statements to the machine that tell it what to do in terms of the hardware componentry of the computer. For example, the "nouns" and "verbs" in the language refer specifically to registers and to the instructions with which the computer is equipped. Languages falling in the third category are called procedure-oriented because the nouns and verbs represent the actions carried out in complex procedures (especially mathematical procedures) and the "objects" handled in these procedures—for example, arrays and vectors, trigonometric functions, iterations of recursive procedures such as the Newton-Raphson method for calculating the square root of a number. Such languages are the principal vehicles for the expression of scientific- and business-oriented programs. In general, the majority of computer programs for control problems are written in assembly language. The second most popular language for the expression of control procedures is the FORTRAN language, one of the oldest scientific languages and the most widely accepted. Languages falling into the second two categories presently require translation before they can be directly interpreted by the central processing unit of a computer, since they are expressed in a grammar, syntax, and symbol set that is not immediately capable of direct processing by hardware. Their form of expression is principally oriented toward ease of expression and reading by the human programmer.

In the following sections we discuss each of these three categories, its characteristics, and the way in which these languages are used.

Machine Language

Of the three categories, machine language represents the lowest level in terms of proximity to the processing unit and distance from the needs of the human programmer. Programs written in machine language are expressed in a strictly numeric character set. In their most general form the character set is that of binary notation: the characters "1" and "0." Each word of the language is a particular instruction to the central processor, formulated in an internal structure such as the one we discussed when we presented the subject of classical computer architecture.

A portion of a program written in a fictitious machine language is illustrated in Figure 4.18. It is clear that it is not easily comprehensible to the average human reader. It is even more difficult to create programs directly in such a language.

One of the first problems encountered in writing programs in machine language is that of keeping track of the locations in which data are stored or to which the program must branch. When the programmer first starts writing down instructions he does not always know the exact address in which the data ultimately will be stored. Furthermore, most programs contain transfers of control, such as the branches and transfer to subroutines that we discussed earlier. It is clear that for large programs (in fact, almost any program longer than thirty or forty instructions), it would be helpful if there were some automatic means of assigning these location addresses and having them placed in the actual instructions. This is the principal function of an assembly program—a translator of assembly language. We discuss these shortly.

Location	Instruction	Meaning (c( ) means "contents of ( )")
0000100000	0010000011001010	Load the A reg. with c(0011001010)
0000100001	0011010011011111	Add c(c(0011011111)) to the A reg. (indirect add)
0000100010	1000001100000101	Shift c(A reg.) left 5 bits
0000100011	0011100001110001	Add c(0001110001) to the B reg.
0000100100	1000001010000010	Shift c(B reg.) right 2 bits
0000100101	1000100000000000	OR c(A reg.) and c(B reg.)
0000100110	0010100001110000	Store result in location 0001110000

. . .

Location	Instruction	Meaning
0001110000	0000000000000000	Location reserved for result
0001110001	0001111111100000	A parameter

. . .

Location	Instruction	Meaning
0010111011	1111111111111011	−5 in binary 2s complement notation
0011001010	0000000000110101	Another parameter
0011011111	0000000010111011	Indirect address constant, points to location 0010111011

Figure 4.18. A portion of a machine language program.

It is clear that the expression of a procedure in machine language is so close to the structural characteristics of the computer that it would be difficult to imagine how any other central processor hardware could be expected to process such a program. The number representing the operation and the address of the operand, for example, must be in particular fields within the individual instruction. An arbitrary alternate computer might not employ the same fields for these instruction parameters, much less interpret the numbers entered in these fields in the same way. Until the advent of microprogramming this was literally the case, and in many ways it is still the case. There are, however, some exceptions. The IBM System/360, for example, can be made to execute programs written for the IBM 1400 computer through the technique of switching to a differently microprogrammed control memory. Technically, the problem still exists since the machine language for the computer is the microprogrammed language. What is actually happening is that the computer is interpreting the binary code in a different way. Operation in this mode is called "emulation," and this capability is increasingly a standard feature of many of the larger computer systems being installed today. It is not yet completely available among the minicomputers being offered today, so the problem of program transferability is unsolved in that area. The principal impetus to the solution of the problem has come from the business and scientific communities where there is a large investment in application programs that are only involved in the processing of data, not the control of equipment and machines.

Ultimately, of course, all programs must find themselves expressed in machine language and they must get to this state by one route or another. The principal vehicles for the translation function are other computer programs that go by the names "assemblers" and "compilers." Assemblers (or assembly programs) are programs that translate other programs written in assembly language into machine language so that they can be directly interpreted by the central processing unit. They are our next topic.

Assembly Language

The first function of the assembly language is to make it possible for the programmer to express his procedures in a character set and syntax that are more easily interpretable by a human reader. To carry out this function another program must be available that translates this language into machine language. This program is an assembly program or an assembler. The programmer expresses his procedure in statements like those illustrated in Figure 4.19. We find essentially the same syntactic structure that we found in machine language and nouns that still refer to specific

```
* THIS LINE IS A COMMENT. USED TO ANNOTATE THE LISTING OF THE PROGRAM,
* IT IS MARKED FOR THE ASSEMBLY PROGRAM WITH A SPECIAL INITIAL CHARACTER
* (IN THE CASE OF THIS EXAMPLE, AN ASTERISK).
START   LDA X1        GET FIRST PARAMETER (THIS IS ALSO A COMMENT)
        ADA X2A,I     ADD SECOND PARAMETER, INDIRECTLY ADDRESSED
        ALS 5         SHIFT SUM LEFT FIVE PLACES
        ADB X3        ADD THIRD PARAMETER INTO B REG.
        BRS 2         SHIFT B REG. RIGHT TWO PLACES
        IOR A,B       FORM INCLUSIVE "OR" OF A AND B REGS.
        STS RSLT      STORE RESULT IN LOCATION "RSLT"
  .
  .
RSLT    BSS 1         RESERVE ONE LOCATION
X3      OCT 17740     THIRD PARAMETER
X2      DEC -5        SECOND PARAMETER, PUT A -5 HERE
X1      OCT 65        FIRST PARAMETER
X2A     DEF X2        PUT ADDRESS OF SECOND PARAMETER HERE
* THIS PROGRAM IS THE ASSEMBLY-LANGUAGE VERSION OF THE MACHINE
* LANGUAGE PROGRAM SHOWN IN FIGURE 4.18
```

Figure 4.19. A portion of an assembly language program.

components of the central processor, but they are now represented by symbols rather than numbers. The function of the assembly program is to read these statements from some input medium, such as punched tape or cards, and to determine automatically the values of the symbols that the programmer uses.

When the assembly program has figured out all of these values, it can then form them into machine language instructions according to the rules established by the design of the central processing unit. Having done this, it then records the instructions on some form of computer-readable material, such as punched tape or cards or magnetic tape. The program may be loaded from this medium into the core memory for execution by the CPU. Programs that handle this loading function are called "loaders."

In general, to formulate completely the machine language program from the statements expressed in assembly language, the assembler must go through the program at least twice. The program in assembly language form can be considered to be made from strings of characters presented in a certain order that follows the rules of the grammar and syntax of the language. In general, assembly languages have a very simple syntactical structure. Statements, for the most part, represent imperatives to the CPU and consist of a mnemonic form of the operations (e.g., CLA for "clear the accumulator") and the operand (a symbolic name for the locations containing the operand or the value of a parameter, such as a shift count). A statement may also bear a label which the assembler interprets as the symbolic name of the location that will contain the assembled instruction. This feature is used to provide names for operands and locations within the program to which the program may branch at some point during its execution. In addition to these the statement may also include various modifiers. For example, the statement might include a modifier which indicates that the operand is to be interpreted as an indirect address, or that an index register is to be used in computing the address when the instruction is executed. During the first pass through the program the assembler formulates what is called a "symbol table"—a list of the symbolic names used in the program and their numeric values. This table is built by processing the instructions one at a time, counting the locations that they occupy, and using this count as the numerical value of symbols appearing in the label fields of the instructions. By the time it has passed completely through the program once, it will have built a symbol table that includes all of the symbols defined by the program. If the programmer used certain symbols that he did not define in some way, then he has made an error and the assembly program will detect this error during the second pass. It is during this second pass that the assembler puts together the assembled instruction by looking up symbol values in the symbol table. If they do not appear in the symbol table, the assembler cannot formulate the instruction in which they are used.

Once the concept of a program preparing another program is adopted, there are many other services that can be performed for the programmer by the assembly program. For example, it is generally useful to break a programming task down into smaller portions, i.e., separate modules or units making up the complete larger program. When the program units are first defined and specified, their exact size is not known and, therefore, the core addresses into which they will be loaded are not precisely known. One service that the assembler can perform is to assemble the program in such a way that the addresses of the symbols are prepared relative to the start of the program unit being assembled. A properly designed loading program can then adjust them when the program unit is actually loaded by adding the first core address to the values of the instruction operand addresses. A program assembled in this fashion is called a "relocatable program," and an assembly program capable of producing such assembled code is called a "relocatable assembler." Incidently, the code produced by the assembler is called "object code," while the original form of the program, as it is input to the assembler, is called the "source code" or the "source program."

Among other services that may be provided is that of automatically generating frequently used sequences of instructions when they are called for by a special instruction in the source code called a "macro." When this mnemonic appears in the input sequence, along with a number of parameters to be filled in as the operands of the instructions, the assembler automatically generates the sequence of instructions associated with the mnemonic and fills in the fields of the instructions with the specified operand addresses. An assembly program capable of handling such functions is called a "macroassembler."

There are many other forms of assembly programs that provide other features of value to the programmer and designer of systems of computer programs. Some, for example, allow the user (within certain limits) to define, by means of special instructions to the assembly program, the way in which the source code is to be assembled. These instructions define the fields of the object code instructions, the values of the mnemonics that identify the operation codes of the computer for which the program is being assembled, and so forth. Such assembly programs are called "meta-assemblers." Other assembly programs provide for conditional assembly, i.e., a process by which the way the source program is assembled may be altered by the values of certain expressions developed during the assembly process. Such assembly features are frequently used in automatic tailoring of programs for varying computer configurations. For example, one of the inputs to the assembly program might be the size of the memory of the computer for which the program is being assembled. Certain instructions in the source program may then be assembled keeping the memory size in mind. Another parameter might be the names of the peripheral devices and the input/output addresses to which they will be assigned. If the device handled by a particular program unit in the source program is not present, the conditional assembly feature may simply pass over this routine when it appears in the source input.

As we pointed out above, most of the programs written for the control of machinery are expressed in assembly language. Computer-aided manufacturing applications display the same characteristics except in those areas devoted to the design area of the manufacturing effort or to the handling of management functions, such as inventory and production control. Programs for these applications are generally written in some form of procedure-oriented language—our next topic.

Procedure-Oriented Languages

Procedure-oriented languages are supposed to be designed primarily for people, although they do make many concessions to the computer. However, they avoid them wherever they can, and research is being done into languages, procedures, and their translation for machines that will someday, if successful, make it possible for the programmer to express his intentions to the computer in carefully phrased English, Russian, French, or whatever his natural language may be. The purpose of the procedure-oriented language is to make it easy to write out the specifications for a procedure; that is, a computer program. The expressions available to the person programming in such a language deal with the terms of the problem more directly than machine and assembly languages. With those languages, the terms were oriented toward the componentry of the computer that was to execute the program (registers, literal core memory addresses, single instructions). With a procedure-oriented language, the programmer writes out the procedural operation directly, using the names of the data and the operations that are to be performed on the data.

In order to express these procedures, the language must contain terms for the expression of arithmetic and mathematical operations, the transfer of data to and from peripheral devices, the arrangement of the data as stored in the core memory or written to the peripheral device, and the way in which the data are to be interpreted. The program, as written in the procedure-oriented language, will be interpreted by another program and translated into machine language so that the CPU can execute the resulting program. Programs that perform this translation function are called "compilers." In general, for each procedure-oriented language/computer pair, there must be a compiler. The compiler reads the program expressed in the procedure-oriented language (the only language that it reads) and translates it into the machine language of the computer (the only language that the computer reads). This translation is not a simple task, and compilers for the most widely used procedure-oriented languages are relatively complex programs.

Figure 4.20 represents a portion of a computer program written in the FORTRAN language, the most popular procedure-oriented language for scientific and control purposes. Its sister language, used in the field of business data processing, is called COBOL (for COmmon Business Oriented Language). The name FORTRAN stands for FORmula TRANslator. There are many minor variations of both languages, so it cannot be guaranteed that a program written in one version will always operate on another arbitrarily selected computer system without some amount of modification.

It is important to recognize that a procedure-oriented language makes certain assumptions about the problems for which programs expressed in this language will be written. For example, the FORTRAN language, being principally intended for scientific computation, provides expressions only for integer and floating point arithmetic. It does not feature any explicit expressions for the manipulation of strings of characters, such as comparing them, inserting substrings, removing substrings, and so forth. This does not mean that the language cannot be used for such functions, it only means that the programmer must use certain ruses to get around the lack of explicit expressions. For example, the language program might treat the characters as 8-bit binary integers with decimal representations, shifting the characters around by multiplying them by the proper powers of two. It is awkward, but it works, and it must consider such matters as the word length of the computer for which the program is being compiled. For example, in a 16-bit computer, only two characters can be stored as a single integer, while in a 32-bit computer as many as four characters can be stored in a single word.

The resulting machine language programs produced by the compiler may not be so efficient as the equivalent program produced by direct coding in assembly language. This is generally because the compiler cannot be provided with all of the information that the programmer might have. It cannot employ clever tricks of coding for the particular problem at hand, and it is also limited in that it must be somewhat general in its translation. Much attention is given to the efficient compilation of arithmetic statements in the FORTRAN language, in which the compiler attempts to retain intermediate results in the computer registers as long as possible, since they might be required in a subsequent operation, and to

```
C A PORTION OF A TYPICAL FORTRAN PROGRAM        Assembly Language Equivalent
C (COMMENT LINES START WITH A "C")
      READ(2,100) PRM                 JSB .READ     Call subroutine to read data
                                                    from device 2
                                      DEF *+5       Subroutine return address
                                      DEF PRM       Address for data storage
100   FORMAT(F3)                      ASC 3,( F3)   Format for reading data
      D=A+B                           LDA A
                                      ADA B
                                      STA D
      E=D                             LDA D
                                      STA E
      F=SIN(PRM)                      JSB .SIN      Call sine subroutine
                                      DEF PRM       Address of sine routine
                                                    parameter
                                      STA F         Store resulting sine
```

Figure 4.20. A portion of a **FORTRAN** program with its assembly language equivalent.

avoid the extra store and retrieve instructions that might be generated. However, for the most part, the machine language code is quite simple and somewhat inefficient. Figure 4.20 shows one such example of this inefficiency, where the sum of the items *A* and *B* are to be stored both in locations *D* and *E*. Note the assembly language equivalent of the code that the compiler would generate. A programmer writing directly in assembly language would omit the instruction that picks up the operand from location *D*, knowing that the same quantity was already in the accumulator of the CPU. Some compilers are "smart" enough to recognize this, but they are generally available only on the larger computer systems.

In many instances, the compiler works by generating "subroutine calls" to already programmed routines that carry out the work implied by the language expression. This is especially true for references to vectors and arrays, and for input/output operations. A subroutine is a special program designed to carry out a predefined function according to certain parameters that are provided to it as input. Generally, the subroutine takes advantage of special subroutine call instructions provided by the CPU instruction set. Such an instruction usually works by storing the address of the location immediately following the location containing the instruction. This address would be automatically stored in some convenient place and the subroutine would then transfer to the address stored there when it finished its work. Thus many other programs could call on the services of this single subroutine, since the subroutine would always return to the caller after it had finished its work. Subroutines generated by the compiler for the purpose of carrying out processing implied by language expressions might be used to obtain a data item from a two-dimensional array, initiate an output operation, compute the sine or cosine of a number, and so on.

What are the advantages of a procedure-oriented language? This is a question of some controversy. The answer is highly dependent on the skills and experience of the people you ask. For example, it is claimed that programs can be written faster in FORTRAN than in assembly language and that they are easier to understand when so expressed. Yet I have seen many FORTRAN programs that were extremely abstruse, and many assembly language programs that bore lucid explanations of what was supposed to be happening, written as commentary along with the code. No one disputes the fact that assembly language is more sparing of core memory and computer time resources. If these are critical resources, the question of language is not even taken up. Generally, if you can spare the core memory, and the problem at hand does not make unreasonable demands on the expression power of the language (i.e., you are able to use the language in a natural way and do not have to contrive complicated expressions to do something that is simple to do in assembly language), then the use of a procedure-oriented language would seem to be the best choice. There are, however, many situations in which this does not apply, and minicomputer systems are frequently an excellent example. Except in unusual circumstances and for fairly simple problems, compilers that operate in minicomputers are not so powerful or efficient as their users might like. They are also rather awkward to use, especially in a minicomputer relying on a Teletypewriter as its principal input/output device. While the program expressed in FORTRAN, for example, may be admirably small in terms of the number of statements of the language, the user will find the following to be frequently the case:

1. Two passes through the compiler are often required; one to get a binary tape of the program in machine language, one to get an assembly language listing of the program so that the programmer can debug the program after it is actually loaded into the computer.

2. The compiler program is read from two punched tapes; one for the first pass and one for the second.

3. Between each pass the compiler must punch a tape that represents the intermediate results of its processing, so that the next pass can carry on.

4. The binary program will be much longer than the original program in terms of the number of characters punched on the output tape.

5. The subroutines that the object program will call are carried on a separate tape, which the loading program must read completely so that all of the called subroutines are made available to the program. This tape is generally enormous, for it must contain all of the arithmetic subroutines, the input/output subroutines, conversion programs, and general utility routines.

The result of all of this is that the user may find himself spending as much as half a day compiling and loading his program, which began with less than one hundred statements in the procedure-oriented language. If the program is not correct and contains bugs, he must temporarily correct them by directly altering the machine language representation of his program in the core memory. When he has this form of his program operating correctly, he must then correct the original version of his program and repeat the whole process, hoping (sometimes in vain) that he has not made another error along the way. You can easily see how a less powerful direct approach might have involved less total work.

Software Systems

We said earlier that the software provided with the computer, and written by its user, served to extend the capabilities of the computer. The prospective user of a computer in the area of direct machine control generally discovers that he has to arrange for the development of his own application programs, either by writing them himself or by subcontracting them to the data processing department within his company, or to one of the many independent software and systems firms that specialize in this business. In either case, the software that the manufacturer of the computer provides with his product is of great importance.

We have already touched on the assembly programs and compilers that make up part of this manufacturer-supplied software. For each computer, an assembler is always provided. For almost every medium- and large-scale computer, one or more procedure-oriented language compilers are provided. However, it is unusual to find efficient and useful procedure-oriented language compilers for minicomputers.

There are other elements of the manufacturer-supplied software that are also of great importance. If programs are to be prepared for and used in a system equipped with punched-tape equipment, editing programs should be provided with the computer so that tapes may be corrected automatically. Loading programs must be provided so that programs can be read into the core memory for execution, and they should be efficient in terms of the core memory that they occupy because they cannot load the user's program into core memory space that they occupy. Most manufacturers also provide subroutines for handling the control and data transfer problems of the peripheral devices that come with the computer configuration. If your application problem involves mathematical work, you should also investigate the mathematical subroutine library that comes with the computer.

Some computer vendors offer elegantly structured and very powerful operating systems with their products. These systems are oriented toward the principal application of the computer. For example, a system intended for stand-alone data processing would provide an operating system that automatically handled the problems of loading the assemblers and compilers when they were required, firing them up, and retaining their outputs. It would then run the compiled or assembled programs and take over the next job down the line when the program was finished. Some operating systems provide for time-shared operation of the system along with the single stream of individual "batch processing" work. This is the type of system we frequently find on large-scale systems handling design automation processing. Some manufacturers also provide operating systems for real-time processing. These systems include the basic routines for recognizing interrupts and calling the interrupt handling programs, handling input/output transfers, and providing for slower response programs to operate in the "background," behind quick response "foreground" processes. These systems should be used whenever they can, since the investment required to duplicate their function is often considerable. The manufacturer has already made a large investment in their development and part of this cost is reflected in the cost of the computer hardware. In some cases, the buyer of the computer may have to pay something extra for the operating system software, and the increased cost may be a worthy investment. However, one should not presuppose that the existence of such a real-time operating system among the vendor's products means that it can, in fact, be used. If it can be used, it should be used, but the matter should be carefully studied before committing oneself to the final decision. One disadvantage of these systems is that they may not be as efficient as a specially programmed operating system. They are generalized and this may interfere with their providing you with such necessities as a sufficiently rapid response time. They may also require more core memory than you can afford, since your own application programs must share the memory with them, and they may include some features that have no value in your particular application. If your real-time foreground and general executive program requirements are relatively straightforward, you may spend as much money tailoring the manufacturer-supplied system to your needs as you would spend directly writing your own, simpler, operating system. These considerations illustrate the fact that, in computer applications as much as in any other field, you should look before you leap.

GENERAL CAPABILITIES AND LIMITATIONS—SUMMARY

There is a great deal more that could be said regarding computer languages and systems, the architecture of both computers and software systems, and so forth. The subject has been treated in an already extensive literature that is growing by leaps and bounds. Many people are saying the same things over and over, but occasionally something new is said. It behooves the person intending to do work in this area, or to use the products of computer science, to try to keep up with what is going on and what is being said. Only then can he evaluate his own needs and their prospects for satisfaction.

One of the first things that must be recognized is that computers are rarely easy to use. There are some problems that can be economically solved only through the application of computer technology. There are also problems that should not be computerized, and some that *cannot* be computerized. The potential application should be carefully studied and the advice of experts should be sought out. Consultants in the computer field are not particularly expensive, and a day or two of their time may save you their fee many times over.

Just as with many other problems, you should know what you are doing and you should work with good materials. Analyze your problem carefully before selecting your hardware and designing your software. Design your system completely before you start to build it. However, all along keep in mind the inherent limitations of the computer and do not allow yourself to be unduly impressed with, or confident of, the ability of the computer to carry out a particular function until you have studied the matter. Keep in mind that the computer is basically a sequential machine (one operation at a time) and that a single computer instruction really does not do very much. Look very carefully at your problem and at the resources that are available for its solution.

We end this discussion of the modern digital computer by briefly touching once more on some of the trends that we have noted at various points in our earlier presentation. The principal current trend is that hardware is becoming less costly and capable of doing its work more quickly. But no great revolutions have occurred. The key word is still evolution and this should continue to be the case for several years to come. Computers get faster, smaller, and less expensive. Peripheral devices get faster and less expensive, but not much smaller. Graphic devices are becoming prominent, largely because they are getting less costly and are becoming easier to use. But the general techniques of applications are not changing very much. If anything, more people now know what the best techniques are and how to apply them. In spite of the well-advertised shortages of engineering and programming personnel, there are more of them in terms of absolute numbers and they are smarter than they used to be. Avail yourself of their services. They are experienced, frequently in problems remarkably similar to your own.

We hope that by now the reader has some appreciation of the characteristics of a modern digital computer, knows some of the terminology, and has some ideas of his own as to how to apply one to his own computer-aided manufacturing problems. Having laid the groundwork for a general idea of the manufacturing process (Chapter 2), the way things are done now (Chapter 3), and the equipment that is available for computerizing these processes (this chapter), we proceed to examine some of the details of actual use of these computers in computer-aided manufacturing applications.

Chapter Five

The Computerization of
Manufacturing Tools

The purpose of this chapter is to give the reader some insight into the technical work required to design and build a computerized electronics manufacturing system. The detailed steps that must be taken are necessarily a function of the specific requirements placed on the system. Therefore, we limit ourselves to the discussion of guidelines, major considerations, and design philosophies. This chapter is principally concerned with technical aspects of the problem, but we occasionally relate these to managerial considerations, which are treated more formally in the chapters that follow.

The company seeking to automate the manufacture of electronic components and equipment has available three routes to the procurement of systems that accomplish this aim. One is through the commercial market in "off-the-shelf" systems. A second is through custom design, fabricating the system from combinations of newly developed components and commercially available components. The third is through the "retrofit" process, adapting existing manufacturing systems to computer control by modifying present controls.

The commercial market in complete off-the-shelf systems is relatively new and the products being offered by companies engaged in this business are just beginning to attain the maturity of established product lines. They are generally represented by small computer systems driving machine tools through specially developed interface systems. There are only a few companies engaged in this market and the potential buyer of these systems has a relatively small range of competing products from which to choose. The technical characteristics of these systems generally adhere to the design philosophies that are discussed in this chapter. This discussion therefore provides some guidelines to follow in the evaluation of commercially available systems, if this is the procurement route chosen.

Most existing computer-aided manufacturing systems were custom designed and installed in the manufacturing facilities of large- and medium-scale corporations. These systems have a conceptual history arising from the individual needs of these corporations. Their development was generally planned by in-house study groups and the fabrication was either subcontracted to outside vendors or handled by in-house engineering groups. In the case of computer manufacturers, the systems were frequently designed to employ the manufacturer's own products.

The third approach, retrofit, is frequently used when the manufacturing facility has many machines already in use, generally operating under conventional numerical controls. The majority of these numerical controls employ punched-tape readers as input devices, and some of the later models provide terminals so that other sources of data may be used as well. In this case, a single computer acts as a substitute for the punched-tape readers and controls a collection of machines. Retrofitting is often the most cost-effective approach to the computerization of an existing installation since new capital investment is minimized while the production of old machines is enhanced.

We begin our discussion of the computerization of electronic manufacturing tools with a treatment of the technical motivations and objectives underlying computerization. If a company is to embark on such a project, it should not be only with a view to keep modern. There should be solid technical and economic reasons for the computerization of a manufacturing facility. Armed with reasons for computerization, we then discuss some general design approaches and philosophies. We finish the discussion with a brief treatment of a fictitious design for a small numerical control system employing a minicomputer as the central control element.

TECHNICAL MOTIVATIONS AND OBJECTIVES IN COMPUTERIZATION

There are many individual motivations behind the computerization of a manufacturing process or any part thereof, but they are all generally derived from the principal motivation of improving the profitability of the operation. If the project does not ultimately result in a more profitable operation, then it should not have been undertaken. We are primarily concerned with the technical aspects of the problem, particularly the questions of general cost-effectiveness and immediate versus deferred benefits.

Among the first things to be considered are the technical requirements of the work that is done in the manufacturing process. As we have stated, the nature of electronics manufacturing is somewhat different from that of conventional manufacturing. This has an important impact on the automation of the electronic manufacturing process, both through the use of conventional numerical controls and through the computerization of the process.

What are the characteristics peculiar to electronics manufacturing? We first notice that the travel distances required of the tools are generally of short to medium dimensions. This is true both for the individual motions in a single cycle of the machine (such as stringing and wrapping a single wire) and for the total dimensions of the completed part (such as a back panel or a printed circuit board). Thus we discover that large machines are not required by the physical nature of the manufactured part. The floor space occupied by these machines is relatively small and they can be placed in close physical proximity.

The nature of the work is also relatively light. The prevalent use of NC machines in electronics manufacturing is for assembly work and light drilling. There is little in the way of heavy metalworking. The parts are relatively light, the largest being wired back panels. Severe electrical and magnetic noise is not usually present and the general nature of their environment is not unpleasant.

Another important feature of the work is that the actual machine assembly operations tend to take place only at the end points of the motion. The cycle is first to position the part, then to attach or insert a component; after which the part is positioned for the start of the next cycle. In fact, outside of metal working for the manufacture of a chassis, the only significant amount of work in which useful operations are performed *along* the path is in the production of artwork using numerically controlled drafting machines. Thus we find that we usually have a situation where a number of small, light machines are employed in very fast work in which terminal position accuracy is more important than motion accuracy. Among other characteristics, we find that there are an unusually large number and variety of individual parts being handled during the manufacturing process.

The complexity and variety of the parts being manufactured create a situation in which the application of numerical controls in some form is frequently attractive from a number of viewpoints. This same complexity and variety of parts often leads to the existence of a large number of separate elements within the manufacturing data base. For many electronic products, especially those destined for military applications or relatively long product runs, a large amount of manufacturing documentation is required. Since there is frequently a parallel engineering effort, engineering changes are another problem with which to contend, in that they contribute to the volatility of the manufacturing data base.

The next step in an examination of technical reasons for the computerization of a manufacturing process is a small study of the ways in which the computer can assist in the solution or minimization of the problems that are peculiar to the electronics manufacturing application. The first technical objective in computerization is the use of these characteristics of the computer to make the manufacturing process more efficient and productive.

To begin with, we take note of the data handling capabilities of the computer. By retaining the manufacturing data base (principally composed of NC programs) in a computer-readable form, one can employ the computer for much of the work of handling the data base and maintaining it in its most updated form. Here, computerization provides a number of advantages over a manual system. Consider, for example, a file consisting of a large number of part programs which are used on several numerically controlled assembly machines for the production of parts, for example, printed circuit boards and wired back panels. When a production run of a particular part is to be made, the part program tapes must be located in the file and taken to the machines to be run. When the production run is completed, the tapes must be returned to the file in the proper place. If more than one machine is to manufacture the part at the same time, more than one copy of the part program tape will be required, since each controller will have a separate punched-tape reader for which a tape will be required. With computer control providing direct drive of the machines, using part programs maintained in computer files, there is no manual handling of punched tapes and no need for multiple copies. By setting up a proper part identification procedure for computer retrieval from the manufacturing data base, selection of the correct part program is assured. Furthermore, loss or damage to the tape is prevented since it is not physically handled.

When the computer is providing direct control of the assembly machines, it will work from a master copy of the part program as it appears in the data base file maintained by the computer system. This provides automatic distribution of any engineering changes to all machines which utilize that part program, since any engineering changes made to the data base are automatically reflected in the master copy. Thus we are relieved of the task of seeing that all engineering changes are reflected in each punched tape and that all obsolete tapes are destroyed or filed where they cannot be accidently used in the manufacture of a part. In fact, the control computer can be employed for the implementation of engineering changes directly to the master manufacturing data base file. This approach provides a system in which there is central authority and responsibility for maintaining the timeliness of the data base and the proper distribution of its components for actual manufacturing operation.

One of the principal concerns in the establishment of a computerized manufacturing system must be the way in which the manufacturing data base is tied in to the automated system in terms of file maintenance and part program distribution. Our concern must also extend to providing some capabilities for the "local" updating of the data base. The purpose here is to maximize convenience by technically permitting some engineering changes and the addition of new parts to take place in close proximity to the manufacturing operation. In this way, the operation can reduce its daily dependence on remote service operations of both the in-house and subcontracted varieties, and thereby improve its response to changes in the product line, emergency situations in product demand, and the correction of engineering and part programming errors. At the same time, proper controls must be established so that unauthorized tampering with the file system cannot take place. The computer can serve as the mechanism for making changes, the communication medium through which changes are distributed, and the element by which changes are controlled.

The speed of the computer can provide valuable services to the manufacturing operation that result in improved productivity, when properly applied. First, the computer's speed can frequently be used to provide a faster rate of manufacture. This improvement, of course, also depends on the mechanical speed of the assembly machines and associated actuators. However, the speeds with which modern digital computers accomplish their processing frequently exceed the operating rates presently available from conventional numerical controls. This is especially noticeable during assembly processes that are essentially fully automatic, such as component insertion and printed circuit board drilling, where the machine cycle constitutes the major portion of the operation. The speed of the computer can also be used to advantage to determine the next step in the assembly process. This step, in a conventional numerical control device, is determined by the data punched into the next block of the NC part program tape. Generally for economic reasons, a conventional numerical control does not feature buffering between the calculation and stepping registers and the punched-tape reader. Thus the controller must wait until it has finished the current operation before it can read the next block from the punched tape and set up its registers for the succeeding operation. The tape reading time contributes a great deal to the overall manufacturing time. This operation need not be so sequential when implemented in a digital computer, for the computer can control the operation of the machine while reading the next block from the data base file in parallel. Thus the data for the succeeding operation are ready and waiting when the current operation has been completed. This is not to say that a conventional numerical controller could not be built in which parallel tape input was part of its original design.

The situation is that most controllers on the market, and in an economic range that qualifies them for consideration as elements of an electronic manufacturing system, do not offer such a feature. It is not required for the majority of conventional numerical control applications (which are in the metalworking industry) where the required rate of work is relatively slow compared to that needed in the manufacture of electronic systems and components. We should also note that the media in a computerized system on which the part program is retained is different and that speed improvements also result from this. Instead of using the conventional punched tape, the part programs would be retained on some form of rotating memory, such as a disk or drum. Such a peripheral device delivers data to the computer at a much faster rate, so that less time is spent in the actual reading of the data.

We have arrived at one of the design features of a computerized system, somewhat by way of the back door, in making our last point about the way in which the speed of the computer system can provide manufacturing rate improvements. This is that part programs are maintained on a rotating (or random access) memory device, instead of on punched tape. The use of such a device is advantageous for reasons other than data transfer speed, as we explain.

Since the speed of the computer generally makes it possible for the computer to control more than one machine at a time, we find that it is desirable to share the processing capabilities of the computer among several machines simultaneously. This has two important possibilities. First, there is the possibility of a smaller investment of capital in the hardware from which the control system is fabricated. Instead of having a separate conventional numerical controller for each machine, we may have a single computer controlling many machines. In the present marketplace, a single conventional numerical control unit costs almost the same amount as a minicomputer with 4096 words of core memory and an ASR-33 Teletypewriter. The principal variation in system cost is in the computer interface equipment, the development of the computer programs that operate the system, and the peripheral memory used to retain the manufacturing data base. For a certain number of machines, the combined cost of these may be less than or equal to the total cost of that number of conventional numerical control units. Beyond this point, the per machine capital cost of the computerized system is less.

The second possibility is that the machines being controlled by the computer do not have to be identical. One might be a semiautomatic wire termination machine, whereas the other might be a printed circuit board drill or an automatic component insertion machine. When we are operating dissimilar machines, we find that the computer must work from more than one part program and that the rate of progress through these part programs will differ from one machine to the next. It is clear that the computer is going to require asynchronous access to different programs. Thus the memory in which part programs are retained must offer random access of one form or another. This is provided by the rotating disk or drum memory. It is true that a similar random access can be made to core memory. A computerized system might therefore rely on the core memory of the computer for the retention of the manufacturing data base. However, a word of core memory is more expensive than the same word available from a rotating peripheral memory. For this reason, part program retention in core memory is used only in systems where the part programs are relatively small and few in number.

The high reliability of the modern digital computer also contributes greatly to improved productivity. The concentration of many manufacturing control functions in a single processor reduces the number of devices in the system that are subject to breakdown or malfunction. The maintenance situation is also improved in that a lower count of system components leads to quicker fault diagnosis and smaller supplies of spare parts. While the modern minicomputer exhibits a remarkably high reliability and short mean time to repair, its peripheral devices do not exhibit comparable figures. Although they are quite reliable, they are frequently the weakest link in the system.

As with almost any good thing, one must always pay something. In the computerized system in which a single computer is controlling a large number of manufacturing machines, the consequences of a computer malfunction have a greater effect. Since there are more machines dependent on the continued operation of the computer for their own operation, a malfunction of the computer system can close down a large section of the manufacturing operation. This is the price we pay and, therefore, one of our goals in the design of a computerized manufacturing system is the minimization of the effect and probability of such a malfunction. Concurrent with this objective is the development of a system in which failure analysis can be promptly made and a repair quickly effected, so that the system can be placed back in full operation as soon as possible.

A manufacturing system must also be subject to the control of the management of the company. To support this objective, one would like to have full (and timely) information on the actual production output of the system and the demands that the manufacturing operation is placing on systems that feed it, such as warehouse systems and outside vendors. The collection of such information is presently attempted in many noncomputerized manufacturing systems by various form and card-oriented production control systems. These systems involve a great deal of paperwork, both on the part of manufacturing personnel and management staff departments. In many companies, the collection of production data, the formulation of these data into reports, and the distribution of these reports to the proper management personnel are the sole functions of relatively large and costly departments. In larger corporations, this information is collected and formed into reports using standard data processing services—frequently the same computer system that is used in the preparation of payrolls, financial statements, and other business data processing. It is frequently the case that the heaviest users of these reports are the middle-level managers directly responsible for the manufacturing operation. Unfortunately, these managers must frequently wait for days (even weeks) for the data to be collected, keypunched, and processed by the corporate business data processing computer.

Given that a computer is directing the operation of the machines doing the actual manufacturing and is working from a manufacturing data base that contains full information on the part in question, could we not reasonably expect the computer to provide quite useful production reports for this management level simply by reporting automatically on its own production output? The answer is yes. There is generally enough processing capacity in the manufacturing control computer to provide production reports directly to the managers responsible for production and to provide these reports much more promptly than the systems currently in use. The incorporation of such a feature in a computerized control system is, therefore, another objective.

So far, we have discovered design objectives in the areas of manufacturing data base management, improvement of the manufacturing rate, sharing of controls to reduce capital costs, failure tolerance, high reliability, short time-to-repair, and provision for management information and control. The next areas to consider are the initial acceptance of the system and the subsequent flexibility of the system. The first few months of operation of a computerized manufacturing system are the most traumatic. It is during this period that the problems of getting used to new ways of doing things must be overcome. Major transitions in the old, familiar ways of doing things are never easy. The transition period for a computer-aided manufacturing system is handicapped by special problems. Principally, these are the logistical and managerial problems of phasing out the old operation and phasing in the new. Rarely are these operations performed by the simple throwing of a switch.

An existing manufacturing system, for example, may already have a large investment in the data base that supports the present product line. If this product line is to be continued under the new system, the data base must be converted for use under the new system. If that is not done, then the new system must be capable of using both the new and old forms of data. This leads us to another design objective: a smooth transition period. For systems being converted from conventional numerical control to direct computer control, this may only involve providing for operations using NC programs punched in the original code. If the system must tie in with a punched-card production or inventory control system, then the transfer of the files used in the old system must be accomplished. This may involve reformatting, alterations in the structure of the files, and the culling of extraneous information. It is a problem that must be studied carefully.

The initial acceptance of the system is also highly dependent on human factors engineering and the philosophies of general operation. The introduction of automation is a touchy subject in some manufacturing organizations, although the electronics industry is relatively free of serious problems in this area. However, the process of familiarizing operating personnel with new equipment is a universal problem. If the equipment is difficult to operate, the performance figures expected of the system may not be realized. If the operating personnel are not happy with the equipment, the general operation will suffer accordingly. Human factors considerations, incidently, extend well beyond the manufacturing operation. The management information system aspects must be properly human engineered as well, and the general aspect of the system should lend itself to presentation as a showcase or upper levels of management may sometimes feel a vague dissatisfaction with the project.

Finally, we come to the question of flexibility. The demands made on a manufacturing operation are generally far from static. They vary with the demands of the marketplace, the availability of raw and semifinished materials, the introduction of new products, and the phasing out of old products. The well-conceived computerized manufacturing system should have an inherent degree of flexibility planned into it, so that it can cope as easily as possible with these fluctuations. The development (or purchase) of properly modular hardware and software can provide some of the required flexibility and, therefore, becomes another design objective. As far as possible, future plans should be factored into the initial design so that, when they are needed, they can be fully and formally incorporated into the system at the lowest possible cost.

GENERAL DESIGN PHILOSOPHIES

There are many approaches to the design of a computer-aided manufacturing system—each having its own merits. In this section we discuss some of these approaches and the principles behind these designs.

The main objective in the selection of a system design is cost-effective improvement of productivity, but this objective is constrained by the availability of the requisite technology and the fund of capital available for the system. There are many ways to attempt this goal, but it is often difficult to quantify the design parameters in a way that makes the most cost-effective approach immediately apparent. For example, one might buy less expensive components with lower life expectancies in the hope that maintenance and replacement costs do not exceed the expense of more costly components having longer life expectancies. This statistical problem is frequently analyzed as a part of the design activity, but the results are often rather murky. There are always enough fortuitous possibilities to make one feel vaguely uncomfortable with the results. If the proposed system involves a large amount of new componentry, then such an analysis can be of some help in selecting this componentry. If this is not the case, then the most cost-effective solution is generally to utilize the more expensive hardware. The best guideline here is the old saying: "You get what you pay for."

Another area in which the factors are difficult to quantify is computer selection. One consideration is the trade-off between the cost of a more sophisticated central processing unit, requiring less programming effort, or a cheaper and less sophisticated central processing unit, more difficult to program and possibly incapable of providing all of the functions planned for the system. Here, one can be trapped by the fact that the cost of the central processing unit is well quantified in the manufacturer's price list. It is very easy to see which one of several competing central processing units is the least expensive. It is much more difficult to see which of the computers will cost less to program and what the difference will actually be.

The quantification considerations emphasize the fact that each system design must finally be judged on the basis of its own merits and requirements. There are, however, several guidelines that can be presented for consideration when a design decision must be made. In the absence of quantified factors that can be subjected to optimizational analysis, these guidelines can be quite useful.

Two aspects of the design of a computer-aided manufacturing system should be considered. The first is the aspect of hardware control and the second is the aspect of the user. A computer-aided manufacturing system is composed of three major parts: the user, the collection of manufacturing tools to be controlled, and the control system. The control system stands between the user and the machines, as a kind of "interface." It is through the control system that the user operates the machinery and it is feedback from the machinery (through the control system) that enables the user to establish and implement the policies under which the machinery will be used. In the sense of design philosophies, this means that the design should be approached from two directions: from the machinery aspect and from the user aspect. These approaches ultimately converge and the result, if the proper decisions have been made, is a properly operating system that fulfills its requirements to the satisfaction of all parties.

In the following material, we are going to treat seven principal areas of design consideration:

1. Analog-digital mix
2. Computer responsibility
3. Data base handling
4. Packaging and general design
5. Human factors
6. Computer requirements
7. Program design

Analog-Digital Mix

We now consider the degree to which analog and digital equipment play a part in the system, with our main emphasis on the selection of actuators. The context is the computerized numerical control system, where the function of the computer is to emulate the operation of a numerical control unit such as those discussed in Chapter 3. This brings us back to the question of using servomotors as opposed to stepping motors in moving the positioning table or the tool head. The reader should refer to that discussion for a review of the relative merits of each (Chapter 3). In examining computerization, there are additional considerations.

In Figure 3.4, a schematic for the numerical control of analog actuators was shown. Analog systems must generally be of the closed-loop variety, since the only reasonably efficient way to control precisely servomotor positioning is through comparison of the actual and desired positions, with the difference controlling the motor drive current. Thus selection of an analog actuation scheme requires the execution of all the functions implied by Figure 3.4: motor drive current control, measurement of actual position, and comparison of actual and commanded positions.

One approach to the computerization of analog positioning is illustrated in Figure 5.1. This is a sampled data system. The computer measures the position of the table (or tool head) and compares it to the desired position. The difference between the positions is used to set a direction and a drive current value which control the direction and speed of the servomotor. The position can be measured by a simple rotary encoding device such as a synchro (which produces a sinusoid related to the angle of the drive

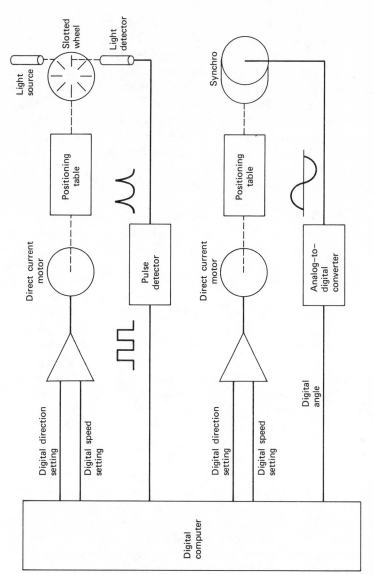

Figure 5.1. Sampled data system.

shaft) or a slotted wheel which exposes a photocell to a light beam. Some rotary encoding devices provide digital outputs that represent the angular position of a shaft. The resulting analog sinusoids (converted to digital values by an analog-to-digital converter) or detected pulses can be used by the computer programs to determine the actual position of the table. A full period of the sinusoid indicates that the table has moved through some linear distance established by the gearing between the table and the encoder. Similarly, a certain number of photocell pulses represents a full turn of the slotted wheel and, therefore, some fixed amount of linear translation. Many rotations of the drive shaft cover the full travel of the table, so the computer must keep track of the number of times that the encoding devices cycle.

The sampled data system is very attractive because very little hardware is needed. The control functions are provided by the computer programs and the interface with the servomotor and positioning table is simple. However, there are some drawbacks. Engaging in a bit of "heuristic" control theory, suppose the bandwidth of the servomotor is F hertz. Precise control of the motors then requires that the computer measure the position of the table and calculate a new drive current value every $1/(2F)$ seconds. Suppose we wish to control N positioning tables, each having two axes. This requires $2N$ control loops and means that the program would (in the worst case) have to perform the control function at a $4FN$-hertz rate. A typical servomotor might have a bandwidth of $F = 25$ hertz. For a 10-machine control situation, the program cycle would be 1000 hertz, or 1 millisecond for the execution of one control operation. A reasonably fast minicomputer might handle this rate; but there would be little or no time left over for processing the other functions required, such as handling management information and the manufacturing data base. If the system could be developed for such a small cost that computerization could be considered for operation of only a few machines, with no major background functions, then such an approach might be the one to take. On the other hand, if the system is to be part of a larger integrated manufacturing system, more of the computer time resources should be reserved for support of functions required by operation in the integrated environment.

A widely different approach is represented by a fully digital system in which the actuators are stepping motors. Here, we can generally dispense with closed-loop control since the distance traveled by the table can be measured by counting the number of index steps used to drive the table. If the machine were to be used for cutting, or milling heavy metal, then there is an appreciable probability that the motor might occasionally fail to step properly. The ability to automatically compensate for missed steps would then be a requirement that would necessitate the presence of a position feedback. But in electronics manufacturing, the heaviest load that the table carries is generally a wired back panel or a stack of printed circuit card blanks. In most cases, as we have noted, the machine is doing no actual work while it is in motion—the actual machine operation action comes at the end of the positioning cycle. There is little reason to expect that loading factors will cause the table to fail to position properly, if the motors have been properly chosen.

A stepping motor is driven by presenting certain binary switching patterns to the motor leads. The fully digital system that corresponds to the analog system of Figure 5.1 is illustrated in Figure 5.2. Here, we have a simple 4-bit output register into which the computer program loads a new binary pattern for each motor step. The rate at which the motor steps is determined by the rate at which the computer loads this register. The direction in which the motor turns is determined by the way in which the bits in the switching pattern change. The counting function, by which the table position is measured, is handled in the computer program. Each change in the switching pattern represents one step closer to the commanded table position, the linear distance covered being determined by the mechanical gearing, the pitch of the lead screw, and so forth.

As with the sampled data analog system, there are certain difficulties

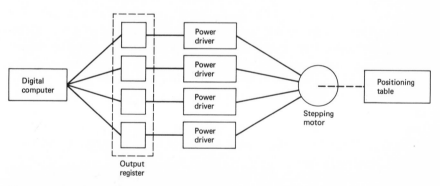

Figure 5.2. Direct stepping motor driving.

with the simple digital system shown in Figure 5.2. Since the rate at which the motor steps matches the rate at which the binary switching pattern is changed, the computer program cycle must match this rate exactly. That is, the motor can only be driven at the rate at which the computer program can change the switching pattern (assuming, of course, that this rate does not exceed the maximum error-free stepping rate permitted by the construction of the motor). Many numerical control systems are geared for linear displacements of 0.001 inch (1 mil) for each step of the motor. Typical table travel speeds attainable by these systems, using conventional stepping motors, are on the order of 100 inches per minute. A little calculation shows that the switching rate required to attain such a travel speed is once every 600 microseconds. Most modern computers have memory cycle times of from 1 to 2 microseconds and the execution of a single instruction generally requires two of these cycles (one to obtain the instruction from the core memory and one to retrieve the operand). Thus between 150 and 300 instructions could be executed during each 600-microsecond period. Again, most computers could handle this rate and control a few positioning axes without too much trouble. However, requiring the computer to operate a large number of axes simultaneously presents problems similar to those encountered in the analog case. If the computer were driving ten machines having two axes each, the worst case situation would allow for only seven to fifteen instruction executions per axis—hardly enough to properly manage the control of that axis.

The two examples just presented are actually somewhat extreme. For example, it would not generally be the case that all of the tables would be operating at their full rates simultaneously. But there are many other considerations that are far from secondary to the fundamental problem of driving the table. For example, processing time must be reserved for accessing the data base, handling miscellaneous functions such as tool head control, wire bin lights, and so on. In the case of the digital system, consideration of the motor start/stop speed must be made. Stepping motors cannot generally be started or stopped at their full stepping rate, so the computer programs must control the rate at which the binary switching patterns are changed when the positioning cycle is starting and as it closes on the commanded position. Starting at too fast a rate can cause the motor to miss steps; cutting down the rate too quickly can cause the motor to take extra steps. With the analog system, overshoot and oscillation or "hunting" must be controlled.

Given all of these considerations, what are the final trade-offs that enable one to decide whether to use digital or analog actuators? The answer is not clear. Consider, for example, two commercially available systems that control semiautomatic wire termination machines. One employs a digital actuator and the other employs an analog actuator. They are comparable in price; but one has not yet demonstrated technical superiority over the other. In one case, high-powered stepping motors are used which have very high start/stop rates. The computer drives the motors only at the start/stop rate and thereby avoids the problem of varying the driving rate. The other system employs logic external to the computer to handle some of the motor control functions, such as comparison of actual and commanded position. This logic is discussed later. It is generally a good approach to the problem of controlling many machines simultaneously in an environment where other processing must be performed.

In the practical case, the selection of the actuator should be technically based on the relative cost of each method, and this can be completely determined only by considering the individual costs of the componentry. Total costs must be considered, maintenance cost as well as original cost. For point-to-point work in which tool head operations occur only at the end points (the most typical case), the relative accuracy of both kinds of actuators is very nearly the same. Servomotors can attain higher top operating speeds. But for short- and medium-travel distances, acceleration can actually be more valuable than rated top speed, as Figure 5.3 illustrates.

Figure 5.3 shows the acceleration, velocity, and distance covered for two different acceleration patterns, a_1 and a_2. The acceleration a_1 is twice as great as a_2, but the actuator with this acceleration has a lower top speed. It reaches this top speed sooner than the other actuator but, because it travels at a higher speed during the earlier portion of the travel cycle, it arrives earlier at certain points. For example, the time to travel distance D'' is T_1 for the first actuator and T_2 for the second actuator. This figure is illustrative of the differences between digital and analog actuators. In fact, the acceleration capabilities of stepping motors are generally greater than those of servomotors. Thus, for moderate distances, stepping motors have higher effective positioning rates. An analysis like that illustrated by Figure 5.3 should be prepared for all of the competing motors that are potentially involved in the system, so that their performance can be compared. If the average distances involved in the work to be performed by the machine are less than the distance D' (the distance at which the higher-speed actuator passes the lower-speed actuator), then the digital actuator has the advantage and should be selected if its cost is within reason.

The last question is the amount of analog equipment to be used in the interface. Here we are considering the possibility of having a separate control unit between the computer and the machine, one which handles the comparison of commanded and present position and the control of the motor. In this case, the computer would be responsible for providing this device with the commanded position. In general, it is better to handle this function with digital logic rather than analog control components. First, the digital logic does not require the digital-to-analog conversion components required by analog controls to convert the digital position provided by the computer into a control voltage. Also, digital logic does not require calibration and is much easier to maintain and diagnose.

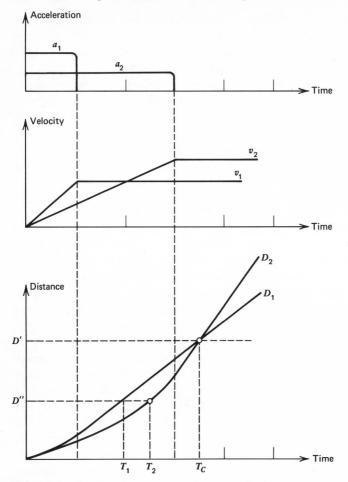

Figure 5.3. Acceleration versus rated top speed.

Computer Responsibility

Having selected actuating elements, the next question is the organization of the position control logic and the general nature of the control interface between the computer and the actuators. The functional requirements of the control problem are, in general, well defined. The most important aspect of the problem at this point, however, is the amount of responsibility that will be allocated to the computer. Some functions may be handled by interface logic standing between the computer and the machinery. We must decide the most cost-effective way of distributing the required functions between the computer and the interface logic.

The basic control problem can be broken down into three basic steps:

1. Measurement of the error between the desired objective and the present state of the controlled object or system
2. The calculation of the control command that directs the object or system toward the desired objective
3. The execution of the control command

Each of these steps has various subsidiary steps and the complete control process must be initialized by injection of the desired objective. In the context of numerical control, the desired objective is the position to which the table is to be moved and is represented either as an absolute position with respect to some origin or as an incremental movement with respect to the present position.

If analog actuators are being used, the present state of the system must be measured by some form of feedback or encoding mechanism. Basically, the system must measure how far the table moved and in what direction. Since analog control is imprecise (due to varying inertias and inaccuracies in the analog control signal), the result of a single control command must be measured after the command has been executed and before the next command can be calculated. With a digital system driving stepping motors, the probability that the actuator precisely obeyed the control command is very high. In this case, the measurement is obtained by assuming that the command was properly obeyed and adjusting the previously known position by the intended effect of the control command—a kind of "implicit" measurement.

In linear positioning systems, the calculation of the control command is generally based on the result of a simple arithmetic comparison of the measured and desired positions. The sign of the result determines the direction in which the table is to be moved and the magnitude of the result determines the magnitude of the control signal and the corresponding rate at which the next stage of the movement will be accomplished. In a point-to-point system, the case of greatest interest in electronic manufacturing, a fast rate of travel is desired when the end point is some distance away. Thus at long distances from the end point, when using an analog actuator, the control signal is generally saturated and the motor is driven at the greatest rate possible. Starting from rest, it goes through an acceleration phase until it reaches its rated top speed. As it nears the end point, the control signal is reduced and the motor enters a deceleration phase. The control signal finally settles into a "dead-band" around the end point. The dead-band is a region centered on the zero error point. Within this region, the control signal is held to zero. The purpose of the dead-band is to prevent minor oscillations, or hunting, in the immediate vicinity of the end point. With a digital system, the stepping motor may be brought up to speed from the starting point by gradually increasing the rate at which the binary patterns are switched, beginning with the natural start/stop rate of the motor. As the table approaches the end point, the control logic gradually reduces the switching rate until the motor's natural start/stop rate is reached. Then, when the end point is finally reached, the pattern is no longer changed. Depending on the relationship between the maximum rate attainable by the motor and the desired top speed of travel, the system may either always operate the motor at the start/stop rate or it may include a rate control algorithm that permits operation at higher rates.

The final step, the execution of the command, requires the conversion of the calculated command into the form required for the actuation of the motor. In the case of the analog actuator, the form is a current that is proportional to the magnitude of the control signal (up to the saturation point of the driving current). In the case of the stepping motor, the form is a sequence of binary numbers that represent the switching patterns which step the motor.

We have examined situations in which computer responsibility for the control of the machinery aspect is at its maximum. These are the analog and digital control approaches outlined when we discussed the analog-digital mix. In both of these situations, the computer handled virtually every step of the control process and was provided with only a rudimentary interface with the machinery being controlled. As we noted, such an approach may not be the most desirable. It may load the computer to the extent that no processing resources remain to be applied to the solution of other problems in the computer-aided manufacturing system, such as data base management and the provision of information to management personnel for the control of the production process.

At the other end of the computer responsibility question, we find some system designs in which the computer assumes a minimum of direct responsibility for the control of machinery and devotes the majority of its processing resources to the data base and management information system. In these systems, we find that the computer is serving as a feeder of control information to devices that are conventional numerical controls, or nearly equivalent. In the conventional numerical control situation, the control information is provided to the controller from a punched-tape reader. In these systems, the punched-tape reader is replaced with a direct link to the computer. Control information (generally NC part programs) are stored on some form of mass memory, such as a disk or drum. The computer retrieves these part programs from the mass memory and feeds the control information contained in these programs to the machine controllers, which in many cases may be conventional numerical control units with their punched tape-readers removed (or placed in a standby position where they can serve as back-up input devices, in the event of a computer malfunction). Many of the commercially available computerized numerical control systems in use for metalworking are variations on this theme, and some of them are combinations of the distributed computer system and the conventional numerical control system.

In these minimum control responsibility systems, the computer is used primarily for management and generation of information and handling the manufacturing data base. The direct control of the machinery is left to the modified NC units. Since the machine control functions can be handled by these NC units, minimum control responsibility systems represent the approach generally taken for the computerization of large existing manufacturing operations. This is the retrofit approach that we mentioned earlier. These are sometimes called BTR systems ("behind the tape reader").

Figure 5.4 illustrates a typical computer-aided manufacturing system of this class. The numerical control units handle the control of the machinery in the conventional ways, the only difference being that direct computer links are used instead of punched-tape readers. Control consoles are associated with each machine for the purpose of selecting part programs and for general communication within the system between operating personnel and supervisors. For example, a machine operator at one of the machines can instruct the computer, via this console, to get a certain part program from the manufacturing data base and begin transmitting that program to the NC unit controlling his machine. He may also instruct the computer to send a message to his supervisor to tell him that work has begun on this part. Similarly, he may receive messages from his supervisor on this console. The consoles may be ordinary Teletype units or small control panels with simple keyboards. Cathode ray tube displays may also be employed for this function.

In this system, the computer is serving principally as a relatively straightforward information retrieval and communication system. This is its principal real-time responsibility. This responsibility does not appreciably load down the computer, so there are plenty of processing resources available for other functions associated with the manufacturing process. Thus the system shown in Figure 5.4 is also provided with terminals for engineering personnel for the preparation of part programs with the assistance of the computer. Here, computer-aided design programs operate in a conversational mode with the engineer and utilize whatever resources remain from the real-time processing functions of part program retrieval and communication with the numerical control units.

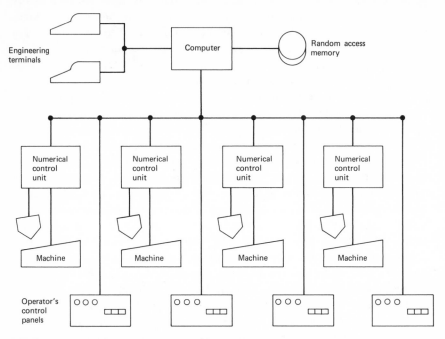

Figure 5.4.   Minimum-control responsibility (retrofit) system.

Between these two extremes of computer responsibility for the direct control of the manufacturing machinery lie many alternative approaches to the problem. We now discuss some of these alternatives.

First of all, there is the question of whether to close the control loop around the computer or to close the loop outside the computer. If the decision is to close the loop outside the computer, one must then decide on the dynamic range of the external control loop. In the case of the retrofitted conventional numerical control system, the dynamic range of the external control system is that of the numerical control unit. By the dynamic range of the external control loop, we mean the magnitude of single-cycle motion of which it is capable. This is determined by the largest number that the control system holding registers will contain and by the way in which the actuators are geared to the positioning table.

If the loop is to be closed outside the computer and conventional controls do not already exist which are to be retrofitted to the computer control, then interface control units will have to be obtained. These interface control units accept control commands from the computer, direct the operation of the machine over a cycle of restricted dynamic range, and then accept the next set of control commands from the computer for the next cycle. They vary in complexity, but they can generally be fabricated from off-the-shelf components. If we are considering a completely digital control system, then we may employ an open-loop control scheme in which there is no direct measurement of the actual movement of the positioning table.

In the two basic analog and digital control designs that we first discussed, comparison between the present and commanded position is handled at all times within the central processing unit by computer programs. In the digital case, the computer program is responsible for determining the next switching pattern to be sent to the stepping motor. This pattern is a function of the last pattern sent and the direction in which the motor is to turn. The computer program can calculate the required pattern, but there are commercially available devices that perform this calculation using wired logic. These devices are called "indexers" or "translators." They are generally available from the manufacturers of stepping motors and are sold as accessories for these motors. There are two inputs to an indexer. The first is a logic level that specifies the direction in which the motor is to step; for example, ground for a clockwise step and negative for a counterclockwise step. The second input to the indexer is either a pulse or a change of level. This input causes the indexer to generate the next pattern in the sequence and to place this pattern on its output terminals. This action causes the stepping motor to take the next step in the commanded direction. In most cases, indexers are implemented as logic devices called "switch-tail ring counters" and are a form of cyclic shift register. Most digital positioning systems employ indexers. The next stage in reducing the computer responsibility for the direct control of the digital system is the use of indexers for the generation of the switching pattern. Such a scheme is illustrated in Figure 5.5.

With the analog actuation system, there is no question of the need for position measurement being directly involved in the calculation of the control command. With the digital system, we may get away with only the feedforward scheme; but direct position measurement can be employed if we wish to establish guaranteed motion or an error-checking capability, although at some additional cost.

Up to this stage of progression in alleviating the direct control responsibility of the computer, all of the arithmetic and rate controlling functions have been retained within the computer. The next step consists of moving these functions into the interface control unit that stands between the computer and the machine. This means that we must have the interface control unit handle counting functions in the digital case and handle comparison and control function calculation in the analog case. When controlling a

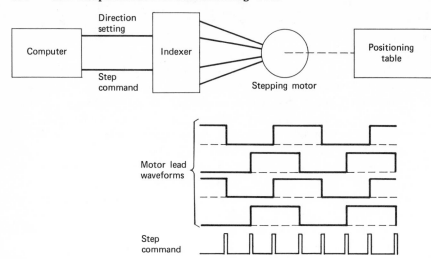

Figure 5.5.   Direct stepping motor driving with an indexer.

stepping motor, the control logic would be implemented entirely with digital electronics. In the analog case, one might employ a mixture of digital and analog electronics. One has the choice of converting the position command from its digital form to an analog form to be used in the comparison of present and commanded position, or one could carry out the comparison function entirely with digital arithmetic logic and obtain a digital command as an intermediate output, which would then be converted to the analog signal that sets the value of the motor drive current.

Figures 5.6 and 5.7 illustrate two such methods of controlling the digital and analog actuators, respectively. The system shown in Figure 5.6 is very similar to the internal organization of a conventional numerical control system for the control of stepping motors, such as that illustrated in Figure 3.3. This system is incremental in nature, with the difference between the present and commanded positions being loaded into the position error register from the computer. The computer also sets a direction flip-flop in the control device that provides the direction level to the indexer. The step command to the indexer is derived from a rate-controlled clock. This same signal is also used to decrement the position error register, providing implicit position measurement.

The system is completely open loop, but a direct position measuring device could be added which would provide a measured position for the computer. This measurement would not participate in the step-by-step control of the motor, but would be used to check the actual motion at the end of the cycle. It would provide a position differential that would be

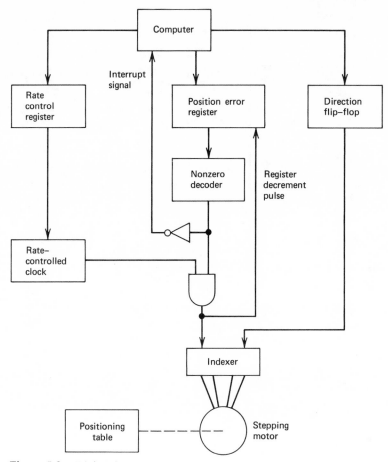

Figure 5.6. Digital incremental computer control.

employed in a small end point correction operation, before the activation of the tool head. It could also be used to monitor the proper operation of the interface control unit and to check on the continuing accuracy of the stepping motor, thereby serving as a maintenance tool. Whether such a position feedback is absolutely necessary is a function of the problem at hand. In general, the added cost would probably serve to disqualify its use, since other means of testing with regular computer maintenance equipment could be employed at significantly lower cost. For example, special NC programs could "exercise" the machine in some form of periodic preventive maintenance program.

If the stepping motors are not to be operated at rates in excess of their

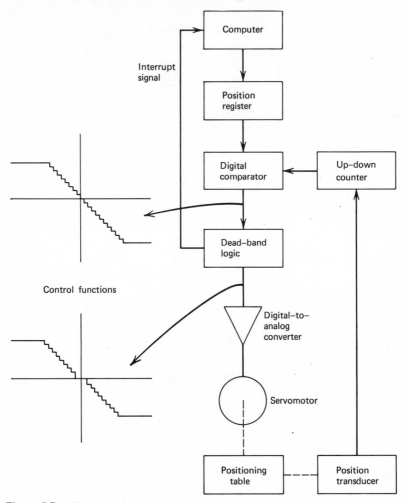

Figure 5.7. Digital position control of an analog actuator.

normal start/stop rates, the rate-controlled clock is not required. However, few stepping motors have high enough start/stop rates for this to be a feasible approach. There are electrohydraulic stepping motors that have start/stop rates in the neighborhood of 2000 steps per second, but they are somewhat more costly than conventional stepping motors. Thus the tradeoff becomes one of more expensive motors versus a rate-controlling device that permits the use of less expensive motors at rates above their start/stop rates.

In the system shown in Figure 5.6, a rate control word is provided by the computer as a digital output to a holding register in the interface control device. The computer is responsible for selecting a rate that will not exceed the error-free stepping rate of the motor, this rate being a function of the present speed of the motor. When a new positioning cycle is starting, the computer program loads the rate control register with the number that corresponds to the start/stop rate of the motor. When the position error counter reaches zero, an interrupt is generated which informs the computer that the last commanded motion is complete. If the motor is stepping at a rate higher than the start/stop rate, the computer must bring the motor down to its start/stop rate before finally bringing it to a stop.

The complete procedure is to break the total motion into a number of smaller motions. The computer would command a start at the normal start/stop rate and would load the position error register with a count large enough to get the motor up to that speed. When this count expires, an interrupt is generated. The computer then loads the rate control register with the next higher allowable rate and loads the position error register with the next increment of motion. This continues until the motor reaches its maximum stepping rate. As the table approaches the commanded position, the process is reversed and the motor is slowed down to its start/stop rate. The last increment of motion would then be allowed to expire normally, and the motor would come to a stop with the table positioned at the commanded point. At the highest rate of stepping, it is imperative that the computer reload the position error register before the time for the motor to take its next step. At high stepping rates, if the clock misses a pulse, the motor may step anyway, since the motor has accumulated a good deal of inertial energy. This step would go uncounted and result in an error at the end point of one or more steps.

In the analog system shown in Figure 5.7, a feedback loop of position measurement is required, since the servomotor is rate controlled and does not respond exactly to command. The positioning command to a servomotor is a signal that sets its rate of turning. In the case of the stepping motor, the command is equivalent to a position command in that it directs the motor to advance one increment of position. A pulse form of position transducer is shown in Figure 5.7. This pulse increments a counter that represents the distance traveled. One might also use any one of a number of other position measurement devices such as synchros or encoded wheels. Generally, the pulse types are either photoelectric devices or devices that employ some form of "fringe" detection, such as counting phase changes caused by the relative movement of two magnetically encoded bars, one attached to the table and the other to the frame.

The device of Figure 5.7 employs digital comparison and control logic to control the servomotor. The reader will recall that we stated earlier that this function might be handled by analog electronics. This is generally true for systems of relatively limited dynamic range and lower precision requirements. The most commonly used control systems now employ digital control logic rather than analog, since the digital logic is more precise and more easily calibrated. It is also usually easier to diagnose and correct malfunctions in digital systems.

In the analog system, the computer loads a position-holding register with the commanded position. This system can be either incremental or absolute and the computer output can accordingly be either the distance that the table is to be moved from its present position or a position with respect to some fixed coordinate system. The digital comparator in the interface control unit compares the measured position with the actual position, providing a position error output that determines the principal part of the control signal that sets the motor rate. Generally, the output of this comparator is proportional to the error, up to some point that represents saturation of the drive signal to the servomotor. If the comparator functions by taking the arithmetic difference of the positions, then the control signal versus error curve is similar to the illustrated curve. Depending on the dynamics of the motors employed, certain simpler comparators could be used. One possible scheme would be to make the comparison on a bit-by-bit basis. The amplitude of the current applied to the motor would be determined by the highest-order bit in which the two positions differed, each succeeding bit of lower magnitude enabling half the current enabled by the next higher magnitude bit. The dead-band area of the curve would be obtained by ignoring the first few low-order bits. Between the saturation region and the dead-band region of the control curve, the signal would be logarithmic in nature.

With schemes of analog control, some method of detecting the attainment of the commanded position is required. This position is attained when the control signal has reached the dead-band region. This condition is used to generate an interrupt, which notifies the computer that the commanded positioning cycle has been completed.

The aspect of dynamic range is very important to the design of these interface control devices. With the retrofitted system, the dynamic range of the system was that of the original numerical control units, the magnitude being determined by the largest number that the control registers in the NC units could hold. This is equivalent to the maximum amount of travel obtainable from a single block command on the part program punched tape. When we discuss dynamic range here, we are talking about the amount of travel that can be obtained through the interface control unit with a single output from the computer. Thus the system illustrated in Figure 5.5 has a dynamic range of only one step, since a single output from the computer causes a table motion of only a single step (in either the positive or the negative direction). With the systems illustrated in Figures 5.6 and 5.7, the dynamic range is determined by the number of stages in the position registers.

The motivation behind the schemes shown in Figures 5.6 and 5.7 is to relieve some of the real-time pressure from the control computer, so that it can attend to other tasks such as data base management and the providing of management information. With the retrofitted system, we have relieved the computer of virtually all of its direct control responsibility, allowing it to serve (from the machinery aspect) only as a substitute for the punched-tape readers that would appear on conventional numerical controls. With the system of Figure 5.5, the computer has full responsibility for all tasks except the trivial one of determining what the next switching pattern will be. The systems of Figures 5.6 and 5.7 lie somewhere in between. But, if the dimensions of the position error registers in the interface control devices are of the same dimensions as those employed in the numerical controls of the retrofitted system, then we have a virtual retrofitted system. Where, therefore, lies the advantage of developing special interface control devices? Why not employ conventional numerical controls, with only the minor modification of replacing the punched-tape readers with computer input terminals? The answer lies in dynamic range; reducing the cost of hardware by employing devices of smaller dynamic range and using the capabilities of the computer to compensate for the smaller dynamic range.

The first point to notice is that conventional numerical control units employ binary coded decimal (BCD) arithmetic. This is because the part program tapes are represented in EIA or ASCII code in which the patterns punched in each frame of the tape correspond to decimal digits or letters of the alphabet. These punched frames are converted to the form used by the controller hardware by stripping off all but the low-order four bits. Thus a register that provides a range of five decimal digits must have four bits for each digit, or twenty bits in all. The counting logic in a stepper motor control and the comparison logic in an analog control do their arithmetic in a decimal base. If the arithmetic were to be done in the computer's natural binary base, there would be less hardware required, since arithmetic in this base is easier to implement with digital logic. Furthermore, fewer register stages are required to contain a number of the same dynamic range, since the binary form of representation is more economical in terms of storage. For example, a register capable of representing 99,999 in binary need have only seventeen bits, a saving of three stages over the BCD representation. In fact, a 17-bit register can hold a positive number up to 131, 071. Thus operating with binary registers and arithmetic can lead to a significant savings in terms of hardware costs. To include binary arithmetic in a conventional numerical controller would require the addition of hardware to convert from the BCD form to binary, an addition that would outweigh the saving from the use of binary operations.

The second point is that we really do not need to reduce the computer load to the extent of having it participate only in the initiation of single block command outputs—the level of the retrofit situation. The main point of providing external controls is to reduce the load only to the point that reasonable resources remain so that data base management and management information functions can be completed in an acceptably short period of time. Once a positioning cycle has been "set up," so to speak, the time required to service an output of one dynamic range is very nearly the same as that required for any other dynamic range. But for each bit that we add to the position error register in the interface control device, we double the amount of time between points at which the immediate services of the computer will be required. The procedure for the computer is simply to take a single command from the part program and break it into a number of individual commands, each one involving a travel distance that is less than the dynamic range of the position error register. Each time the zero point is reached, whether by counting motor steps or attaining the dead-band region, the computer loads the position error register with the value for the next step in the cycle.

For example, suppose that we have a part program command which requires a travel of 1000 steps in some coordinate direction. Suppose further that it takes the computer N microseconds to recognize that an output is called for (recognize an interrupt) and to output the next value to the position error register. With the simple interface device of Figure 5.5, $1000N$ microseconds of computer time would be required during the travel of the table. If the motor stepping rate is 1000 steps per second (at 1 mil gearing, a travel rate of 60 inches per minute), then only $10^3 - N$ milliseconds of free time would be provided. On typical minicomputers, N can be as much as 100 to 200. Thus motor service time represents from 10 to 20 percent of the available time. When we add the fact that quite a few axes must have the same attention paid to them, we have a rather heavy load.

Now suppose we are employing an interface control device with a position error register having four bits and therefore capable of handling a count up to $2^4 - 1$, or 15. The computer could then break the 1000 step motion in sixty-six increments of fifteen counts each, and one final increment of ten counts. The interrupt service time can be assumed still to be very nearly N microseconds, so that now we find the computer spending $67N$ microseconds out of the one second travel time. Thus $10^3 - 0.067N$ milliseconds of free time are provided. Here, the computer would be devoting approximately one percent of its time to the direct control of this axis. So you see, very little in the way of dynamic range in the interface control device provides rather spectacular time savings in the computer. And for each bit that we add to the position error register, we halve the amount of time required for direct service of that axis.

The analysis of the number of bits needed in the position error register may be roughly parameterized as follows. Let

A = the number of axes to be controlled,
S = the amount of time required to service one axis,
B = the number of seconds per second to be allocated for management of the data base and management information functions (e.g., 50% allocation implies $B = 0.5$),
R = the maximum motor stepping rate (steps per second).

Then the amount of time allocated per second to the control of each axis is given by $(1 - B)/A$ and the number of times per second that the axis can be serviced is given by $(1 - B)/(AS)$. In one second a motor operating at top rate will cover R steps. Thus the number of steps per service is given by $(RAS)/(1 - B)$. The register should be large enough to hold this number of steps. The number of stages required for a register this size is

$$m = 1 + \text{greatest integer in } \left[\log_2 \left(\frac{RAS}{1 - B} \right) \right].$$ (5.1)

Incidently, a restriction on the value of S is

$$S \le \frac{1}{RA}.$$ (5.2)

Otherwise, if all axes request service simultaneously, the time between the request and service of the axis with the lowest priority will exceed the interstep time. If the motor is operating above its start/stop rate, overstepping may result.

To summarize the guidelines that we have developed so far, we find that we will be either retrofitting the computer control to existing conventional numerical controls or we will be designing and building special interface control devices. In the case of retrofitting, the modifications required to the numerical controls are relatively minor and they can generally be driven from the computer using more or less standard peripheral interface techniques. In some cases, the manufacturing system can be spread out over a rather large floor area. The machines and the computer may actually be in different buildings. In this circumstance, conventional communication techniques can be employed which use serial transmission technology with high-noise immunity. Interface devices are in standard use for interfacing computers with telephone lines and these same devices can be employed for the purpose of communicating between separated computers and manufacturing equipment.

If the system does not involve retrofitting, then special interface devices must be obtained. Generally, these are somewhat inconvenient to use when long transmission distances are involved, since they require rapid response times which are rather difficult to attain with serial communication techniques. However, they do represent less hardware cost than conventional numerical controls. This reduction in cost is obtained through the use of binary counting techniques and reduced dynamic ranges. In laying out the design for one of these devices, one first determines the required size of the position error registers. This requires an analysis of the computer time resources needed to handle the data base management and management information functions. Whatever time is left may be devoted to direct machine control and distributed over the number of machines to be controlled by the system. The number of bits provided in the position error registers should be adequate to provide this time.

Data Base Handling

In addition to the function of direct machine control, the computer-aided manufacturing system also must assume responsibilities for the management of certain data associated with the manufacturing process. The way in which these responsibilities are carried out can have an important effect on the overall performance of the system. It is therefore important that they be properly considered in the initial design of the system. From a hardware point of view, adequate data storage facilities must be provided and they must offer sufficiently rapid access to the data, so that the control and reporting functions can be handled in the requisite time. From the software point of view, the data files must be structured in ways that allow the computer programs to create, retrieve, and modify data quickly and efficiently. Each aspect of the manufacturing problem influences the way in which data pertinent to that aspect should be stored and handled.

To begin with, one should examine the characteristics of the data used in the computer-aided manufacturing operation. These characteristics determine the proper disposition of the data. Data that cannot be made to "flow" are of no value, so another important set of attributes that must be considered are the relationships between the sources of data and the various users of the data. In addition to the data base (the data that specifies how a particular part is to be made), there are other forms of data that must be moved through the complete system. These are such items as production orders, purchase orders, reports to management on inventory, production output, and so forth. There is generally no need to build large files of these items, although they do have a short, but finite, life in the system. The manufacturing data base consists principally of part programs and, occasionally, bills of material and part explosions; beyond this, much of the information is devoted to management functions— the control of the output of the production system, as opposed to the direct control of the machines themselves.

The file of part programs is common to virtually every computer-aided manufacturing system. The information that is contained in this file is nearly the same information that is contained on the punched tapes used in conventional numerical control systems. For each part, there is a part program. This program consists of individual blocks of data, each block being formulated according to a particular format. In conventional numerical control systems, the format is determined by the NC unit that is being used. The Electronic Industries Association (EIA) has established a set of standards for the representation of numerical control programs and most NC systems employ one or more of these standard formats. The codes employed for the individual characters that make up each block may be either the codes established by the EIA standards or they may be the ASCII codes that are commonly used for information interchange between computers and communication systems.

When a computer-aided manufacturing system is ordered to produce a certain part on a certain machine, it must first obtain the part program for that part. Generally, the part programs are contained in some kind of random access memory, such as a disk or drum. Also contained in the peripheral device is a directory of the part programs. This is a separate file that contains code numbers of the characters that form the name of the part program and pointers to the program. These pointers are representations of the address of the part program on the random access memory. Recall from Chapter 4 that disks and drums are divided into tracks and each track is divided into a number of individually addressable sectors. Thus the pointer for a particular part program might consist of the track and sector at which the part program begins. When ordered to manufacture a certain part, the computer locates the part program by looking it up in the directory and retrieves the blocks that specify the individual steps in the part program from the memory. Part programs frequently consist of many thousands of characters. In most computer systems there is not room to store the complete part program in the core memory along with the computer programs that direct the operation of the system. Thus the control computer may retrieve only small portions of the part program at a time.

Since the control computer is generally responsible for the control of several machines, the times at which the computer will wish to gain access to the part program file for the retrieval of the next block will come at random and there will be situations in which different blocks from the same part program are needed for more than one machine simultaneously. Thus one of the major requirements on the part program file is that it must be amenable to random access. It is also important to note that this access is of the real-time form. That is, the retrieval of the next block or group of blocks must take place in the smallest possible time, since a manufacturing machine is waiting for the data contained in the block.

Another important characteristic of the part program file is the possible volatility of the information that it contains. Although there are situations in which the part program file is rather static, it is most generally the case that part programs are often changed. Engineering changes to established product lines are a major source of part program volatility. In some manufacturing organizations, a kind of "job shop" operation is employed, and the parts being manufactured change from one contract to another.

There is a whole class of computer controlled NC systems in which the part program file is the only file of any consequence. These are strictly direct control sysems with no links to any element of the corporation off the manufacturing floor. But, when a computer-aided manufacturing system develops such links, other data files can come to be of great importance to the system. One such file is an inventory file. In the system, raw material is being converted to finished and semifinished parts by the manufacturing operation. The effect of this on the inventory is to convert certain part numbers into other part numbers and to modify the on-hand quantities of parts. In practice, for example, the automatic insertion of components and DIPs in a printed circuit card reduces the count of on-hand components and DIPs and increases the count of on-hand completed printed circuit cards that are, in turn, separate subassemblies in a larger electronic product.

In the complete computer-aided manufacturing system, each step of the manufacturing cycle can update the inventory file automatically. To perform this function, the computer must have access to the inventory file. The frequency of access to this file is much lower than that for the part program file since the data change only once per part program run. Another characteristic is that only a single access is required to update the file. This would be done at the end of each machine's manufacturing cycle. It is also the case that the information content of the changes is relatively small, since it represents primarily on-hand count changes. Here we have a situation in which this information might conceivably be collected over some period of time, after which the modifications to the file would be made in a batch. This would reduce the frequency of access still more. The file could be updated directly by retaining the file on a peripheral device attached directly to the control computer. If the file were not directly accessible to the control computer, then it could prepare a report delineating the inventory file changes. This report could then be employed to make the actual changes in the file using either manual methods or a separate computer system, such as an ordinary business data processing computer.

To give the control computer the capability of updating inventory according to actual manufacturing operations, the computer must have supplementary information about the individual components that make up a complete part or subassembly. Since the part program represents the way in which individual parts are transformed by the manufacturing operation, it is with the part program that this information is most naturally associated. Each time a part program is run on a manufacturing machine, it has a particular and well-defined effect on the inventory. Thus in the minimum system the control computer's report would consist of a count of the number of times each part program was run during the reporting period. In the more advanced system, each time the part program was operated, the supplementary information would be used to update the inventory file directly. This supplementary information consists of a list of the part numbers that make up the part being manufactured. This information is the principal ingredient in a "part explosion" or "bill of materials" file, a file that lists, in some useful order, the component parts of each subassenbly.

The part explosion file is both a part of the manufacturing data base and a link to the inventory system. It can also serve as a control link on the actual manufacturing operation. For example, if the system is ordered to produce a certain printed circuit card, a number of cycles of an automatic component insertion machine must be run. During each cycle, components of a particular type are inserted in the board, the components being obtained from a presequenced roll. For each component, there is a part program that specifies the table positions at which the component is inserted. Given a properly conceived part explosion file, the control computer can reference the inventory file to see that enough components of each type are on hand to complete the manufacturing run. As each cycle is completed, it can update the inventory file accordingly or prepare a summary report. Here we find a method of access and frequency of access that are comparable to those required for the inventory file. Retention on a random access memory device is again indicated. If prechecking of the inventory before the start of a manufacturing operation is desired, then the computer must have direct access to the inventory files. Otherwise, the computer can prepare summary reports that are used for later file updating. If the files are retained on different computer or file systems, then procedures are needed for coordinating the part explosion file with the part program file that the manufacturing computer is using.

Beyond the maintenance of inventory files, the computer can also formulate useful reports to the management of the manufacturing organization and allied departments. Direct control of the machinery, for example, gives the computer the opportunity to measure directly the performance and production output of each machine under its control and to formulate this information into printed reports. Its handling of the inventory system can provide information on the state of the inventory, such as potential shortages of components. It can automatically initiate replacement procedures for the components that it draws from stock for the manufacturing operation, when these fall below certain critical levels. The majority of the information that would be placed in management reports is derived from monitoring the daily operation of the manufacturing system. Since the computer is at the center of this activity, it can provide this information as a relatively straightforward by-product of its normal manufacturing functions.

The characteristics of access, modification, retrieval, and volume for each major class of data to be handled by the computer-aided manufacturing system must be carefully analyzed when designing the system. In smaller systems in which direct control of machines is the only principal function of the system, one may find that the desired system, or a reasonable approximation, is commercially available. With such a system, a method of access to the part program is provided and the volume that the file can attain is one of the parameters of the system, with larger volumes occasionally being available as extra cost options. Even in this case, the buyer should carefully analyze his own requirements to be sure that the system will satisfy them. Thus careful attention should be given in both cases to the number of part programs that need to be maintained in the system, the size of these programs, the frequency with which they will be changed, and the methods provided for modifying programs in the file, for deleting old programs from the file, and for adding new programs to the file.

The volume occupied by a part program can sometimes be surprising. For example, in conventional NC form according to most of the EIA standards, a single block of word-address formatted NC program for a two-axis machine can consist of as many as twenty-seven characters. If the machine is a semiautomatic wire termination machine, two blocks are required for each wire. Thus a part program for a back panel of 1000 wires would require 54,000 characters. In a 16-bit computer, storing the characters in unmodified form would require 27,000 words of storage. This is an appreciable amount of storage for only a single element in the data base. Fortunately, several methods exist for compacting this information into less bulky forms. One method often used consists of retaining the information in a fixed format in which sequence numbers are omitted, the positions are converted to their binary representations, and the miscellaneous function codes are converted to binary and packed several to a single word. With this technique, the single block can generally be compressed into from three to five or six computer words, depending on the dynamic range of the position data and the number of miscellaneous functions required by the machine. These techniques should be employed wherever they can reduce the storage volume requirements of the system.

Just as one may discover that what seems to be an enormous amount of peripheral memory is, in fact, somewhat short of one's expectations, one may also be surprised at the amount of time it takes to transfer data from its original form to its place of normal use on the peripheral random access memory. When using the tape reader of an ASR 33 Teletypewriter, 90 minutes are required to read the 1000-wire example program into the computer. With a typical high-speed punched-tape reader, the reading time is 3 minutes. One should therefore generally plan on loading these data files with the fastest input devices that one can reasonably afford.

The techniques by which the files are accessed and retrieved for reference and modification are also very important. Directories should be employed for the initial location of part programs in the manufacturing data base. Techniques should be used that permit the storage of part programs of varying lengths in an economical fashion. For example, the most efficient means of storing a part program for fast retrieval is to store all of the blocks sequentially. As a new program is added to the file, it is stored beginning with the next free storage area following the last program added to the file; and its starting location is entered in the directory, so that the computer can locate the part program quickly. If, however, a program of a certain length is deleted from the file, then the space it occupied becomes available for a new program. If the new program is added to the file, it should fill in this area. If the new program is shorter than the deleted program, there will be some space left over which may be too short to hold a subsequent complete new program. If this process of deleting and adding part programs is carried out several times, the file may end up with a good deal of empty space. The total volume of this space may be useful, but it is broken up into a number of smaller segments that are individually unusable. In this circumstance, revision of the complete file is called for. During this revision, the part programs would be "packed" back into the storage area to eliminate the small free segments and leave a large free segment at the end of the file area. Computer programs to perform this function of "file cleanup" should be provided with the system, if the part programs are to be stored in the sequential fashion.

Another method of file storage is to divide the complete file storage area into blocks of a fixed size with pointers included at the end of each block. These pointers contain the address of the next sequential block in the part program and serve to connect the blocks in the logical sequence of progression through the part program. This technique is called file

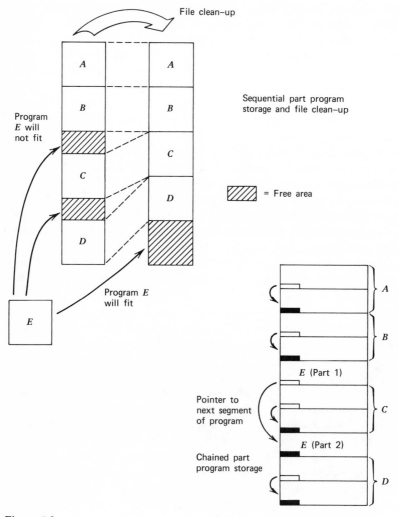

Figure 5.8. File storage techniques.

"chaining." It has the advantage of reducing the need for periodically going through a file cleanup process, but it may result in less than optimum speeds in accessing the files. Figure 5.8 illustrates both the sequential and chained techniques of file storage.

Packaging and General Design

The packaging and general design of the computer-aided manufacturing system hardware principally influences system flexibility, reliability, and ease of maintenance. The key design consideration here is "modularity." As far as possible, the system should be fabricated from independent functionally oriented subassemblies (modules). This consideration extends even to the system software.

An important consideration in the selection of a computer for the system is the modularity of its design and packaging. In many recently introduced minicomputers, the current trend is to package the central processing unit on a separate printed circuit board. The core memory is also packaged separately and the peripheral device interface logic is either mounted on sections of the same large board or each device is allocated a separate card. In the case of some minicomputers, the complete package is often small enough and of such low cost that the whole assembly could be economically spared and entirely replaced in the event of a malfunction. If the system is more ambitious in scale, such as a fully integrated manufacturing system that handles inventory and production schedules, as well as direct control, the system design may include several computers. The larger computer would handle file control and management information and supervise several minicomputers which would be responsible for the direct control of the machinery. In a system of such scale, the concept of complete spare minicomputers is quite attractive. In smaller systems, the ratio of spare parts cost to the cost of the system would be much larger if the whole control computer were spared, and might not be generally cost effective. In such a case, the modularity of the computer is then of greater concern.

The modularity of the interface control devices must also receive careful attention. Here we have a situation where, using modern integrated circuit and packaging techniques, a control interface device could easily be packaged on a single card, or at most a few cards. By packaging it as a single mechanical assembly, the level of technical proficiency required to effect quick repair of a malfunctioned system is significantly reduced. The control of the machine tool also requires certain miscellaneous functions which are generally simple discrete signals that drive (directly or indirectly) relays in the machine and indicators. These should be generalized as much as possible and packaged separately from the control interface devices, but should be placed in the equipment racks in close proximity to the control devices. This is to facilitate quick fault location by technicians with low training levels. In most systems, for example, it is feasible to provide two simple mechanical assemblies for each machine. One would contain the control registers, clocks and counters for all axes of the machine. The other would contain the discrete functions, either relays or relay drivers.

Figure 5.9 illustrates the general concept of a control system with the kind of modularity that we have been discussing. The details of the inter-connection between the interface rack and the computer depend on the input/output structure of the computer. For some computers, there might be a separate input/output channel between each module in the interface rack and the computer. In others, additional logic might be employed to gate the various devices in the rack onto the computer's input/output lines according to device select codes provided by the computer.

There is considerable flexibility with regard to the actual location for mounting indexing logic (indexers) if stepping motor controls are employed. In most cases, stepping motors require four leads carrying 40 volts or more. The inputs to the indexing logic are generally logic voltage levels and consist of two lines. There are several good arguments for mounting the indexers near the machines, rather than in the interface rack. This requires only two logic level lines between the interface rack and the machines. The principal problem is the maintenance of stepping command integrity, particularly if the indexer steps on transitions with critical tolerances. The alternative is to mount the indexing logic in the interface rack and carry out the four higher voltage leads to the motors. The final decision should be determined by the influence of noise on the stepping command signal and the relative costs of cabling for the two alternatives.

The majority of the logic and mechanical elements required for the control devices in the computer-aided manufacturing system are relatively simple and quite reliable. One important challenge to the designer of the system is to assemble them in a way that will be amenable to quick fault location and simple prompt repair. The reliability of the computer is also generally high. It is not unusual to find modern minicomputers with mean times between failure of one year or more, even under three shift operation. The system area in which reliability problems are most likely to occur is the area of computer peripheral devices. Severe mechanical demands are made on these devices and one must expect and prepare for the prob-lems that will undoubtedly occur. ASR33 Teletypewriters, although they are the most frequently used computer input/output device, were not in-tended for the heavy-duty use to which they may be subjected in computer-aided manufacturing applications. Heavy-duty versions are available as

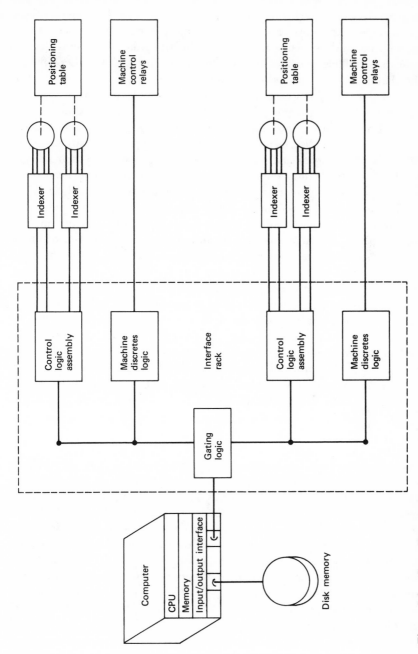

Figure 5.9. General control system organization.

extra cost options which should generally be elected, since they may pay for themselves very quickly in reduced maintenance costs. High-speed photoelectric tape readers are usually quite reliable, since they are very simple mechanically and are self-synchronizing (the synchronization being derived from the feedhole punched in the tape). High-speed punches, on the other hand, are relatively unreliable and require frequent preventive maintenance. They are, however, a necessity if the system is to be used for the creation of part programs in any large volume. They are also valuable for the production of hard copies of versions of part programs modified on the system. The disk memories used for retention of the part programs are highly critical items and great attention should be paid to their reliability. Head per track disks offer greater reliability than moving head disks, but are somewhat more costly in terms of unit storage per dollar. The principal problem with the moving head disk is the head. The rapid, short, and precisely controlled motion of the head is difficult to maintain over long periods of use. An alternative for installations that expect to maintain large files of part programs is to have both a fixed and a moving head disk, the former of restricted capacity. The fixed head disk is used as a buffer storage and the moving head disk is used for the permanent storage of a part program file. Thus when a production cycle on a certain part begins, the computer makes a single access to the moving head disk to retrieve the complete program and transfer it to the fixed head disk. The fixed head disk contains only those part programs for parts under active manufacture. This provides large storage capacity, but avoids the heavy use that creates reliability problems for the moving head disk. It is also less costly than a large amount of fixed head disk storage capacity.

Human Factors

Human factors play an extremely important role in the long term success of a computer-aided manufacturing system. Acceptance of the system by the people who must deal with it on a day-to-day basis will be principally determined by how "comfortable" they feel with it. By comfortable, we mean the feeling that the computer system is *assisting* in the performance of familiar tasks, enabling people to perform their work in a more relaxed manner. To achieve this, the system must display "tact." For example, from the human factors viewpoint, it is better if the presence of the computers prevents errors from occurring in the first place. If the computer is there to catch and report errors to supervisors, the worker may feel threatened rather than helped.

Closely related to the idea of the computer helping (rather than controlling) the work activities of the manufacturing personnel, is the question of the ease with which the computer is operated. In a proper computer-aided manufacturing system, the control computers should be inconspicuous and maintain a "low profile." The only times at which operating personnel should have direct communication with the computer is at the start of new production cycles, when new part programs must be assigned to certain machines. This is generally a supervisory function in an assembly line or product manufacturing operation, but the individual worker can participate in this function in manufacturing organizations that operate on a "job shop" principle.

There are two points at the machinery end of the system at which operator-computer communication takes place directly. One is at the semiautomatic machine where, at the end of each step performed by the operator, the computer must be signaled to proceed with its next step—generally, the table or head positioning cycle. Here the mechanism is usually a simple foot switch or button. The interface is simple and differs not at all from the interface that the operator had with the conventional numerical control unit. The computer introduces no new training problems at this point and may, in fact, occasionally assist in this function. For example, additional discrete indicators can be provided and driven by the computer to assist in the training of new operators. The computer can be set up to provide special training part programs that acquaint the new operator with the features of the machine.

The other point at which direct operator-computer communication takes place is the control console of the computer. One should distinguish between the front panel of the computer CPU and the device used to control the computer programs. Generally, the device used for this latter function is a Teletypewriter. The Teletypewriter keyboard is used to enter commands and the associated printer is used to record the responses and acknowledgments of the computer. The use of a Teletypewriter is the lowest cost solution to the computer control requirement for small systems. Some systems may require special control consoles or panels, especially those in a job shop operating environment. In such an environment, each machine operator has to communicate with the computer in order to select the next part program for his machine. The Teletypewriter may not represent a cost-effective solution to this requirement. Generally, the better solution is a collection of simple keyboard devices for command input with indicator lights for presenting the computer's response to the operator. It is most often the case that these consoles must be tailored to the procedural requirements of the machine to which the console is assigned. For example, with semiautomatic wire termination machines, it may sometimes be useful to display the current line of the wire list being processed. Cathode ray tube display devices can be found useful for these applications, but they are somewhat more expensive than Teletypewriters. They are, however, more reliable, require less maintenance, and use no expendable supplies.

The design of the keyboards used on these control devices can be oriented toward operation by personnel with relatively good manual skills, but not necessarily approaching those of a typist. In most cases, production-line personnel in electronic manufacturing operations have reasonably high degrees of manual dexterity and good visual responses, so the operation of most keyboard devices presents no special problems to them. In special circumstances, some commercially available keyboard/CRT displays can be employed for use by personnel with lower levels of manual skills. These devices arrange their keyboards in numerical and alphabetical order so that the operators can locate the required keys more easily.

If the control requirements of the system can be satisfied by the Teletypewriter, then the human factors problems focus on the software and the formats that the software employs for input commands and printed responses. Programmers are notoriously prone to abstruse and overabbreviated printouts and commands, and the formats they select are occasionally awkward to use. This aspect of the system should be carefully specified if a new design is being used. If a commercially available system is being procured, samples of typical control sequences and printouts should be obtained from the vendor for study with regard to their satisfying the requirements of good human factors considerations. Generally, the best formats from the human factors point of view are "free-field" formats. The closest analogue to this, which may be familiar to those with training in conventional numerical control systems, is the word-address format for part programs. Here, each control parameter is preceeded with a letter that identifies the parameter. A similar form of coding for computer command inputs can generally be adopted for the computer control system. Another variation of the free-field format is one in which the position of the parameter within the input command string identifies the parameter. In this free-field format, the individual parameters are separated by some common punctuation character, such as a comma or slash. The numerical control analogue to this format is the tab sequential format, in which parameters are separated by the tab character. If a parameter is omitted from the sequence or intended to be zero, two sequential punctuation characters are used.

Operation of the system on a day-to-day basis usually requires no training of personnel in computer programming. Programs are generally kept in the core memory or on a disk. Operating system software handles any program changing functions, using directions provided from the control console. There is little in the way of system operating characteristics to indicate to operating personnel that a computer is actually involved in the operation. In fact, there are some arguments for not identifying the system to its operating personnel as being computer controlled. It is simply a piece of control hardware that performs certain useful functions in making their work easier, more efficient, and more productive. The personnel who will require the greatest degree of training will be maintenance personnel. Electronic technicians of average skills can handle most maintenance problems and every computer manufacturer offers extensive training programs at modest cost.

If the system involves management information and the preparation of part programs, then those personnel who employ these functions require additional training. In most cases, the operation of the management information functions are quite simple and consist mostly of uncomplicated orders to the computer to produce a certain report. The format and content of the report has been determined as part of the system design and, as these reports generally require little in the way of control parameters, they proceed rather naturally and independently. The manager's choice in requesting the report is usually limited to the specification of a format. That is, he may request an abbreviated report, a full report, or a summary.

The operation of the engineering functions falls under the category of computer-aided design. Operating the system for these purposes can be rather complex, but no more so that the equivalent procedures on a stand-alone computer system or a time-sharing system. In fact, the use of the computer for these functions may be identical to that of those operating regimes. For example, the computer system might have a Teletypewriter in the engineer's office which he uses just as he would use a time-sharing terminal. Alternatively, the computer can run computer-aided design programs in a batch-processing mode during the hours that the manufacturing operation is off.

Our last comment on the human factors subject has to do with the general presentation and aspect of the system. Straightforward functional operation is a good feature, but it can be appreciated in value considerably by the general aesthetic aspect of the system. Displays are easier to read if they are presented in ways that are attractive to the eye. The general feeling toward the system is also influenced by the colors that are used on the control panels and racks. Warm, friendly, colors have an excellent effect on operating personnel.

Computer Requirements

The general characteristics and parameters of the computer required for the computer-aided manufacturing system are determined by a number of factors, among them the following.

1. The number of machines to be controlled
2. The speed of the machines controlled
3. The number and size of the active part programs in the manufacturing system
4. The degree of expansion flexibility that is to be planned into the system.

Insofar as it is possible, given the temporal and economic constraints of the development schedule, the system requirements should be laid out in some detail before the matter of computer selection is finally settled. The principal parameters in computer selection are price, speed requirements, core memory requirements, and peripheral device requirements. The design of the computer system will influence these items to the extent that it makes equivalent programs longer or shorter, both in terms of execution time and the number of instructions required to perform certain functions

A good central processing unit instruction repertoire reduces the effort required for programming to some extent, but the reduction is rarely measureable. The greatest advantage bought with a good instruction repertoire is generally smaller and faster running programs. This can result in smaller core storage requirements and the possibility of employing a slower computer and, therefore, a less expensive one. Computers with larger word sizes usually have the advantage of a better instruction repertoire on larger pages of memory (see Chapter 4), both of which result in faster and smaller programs. A more complete instruction set enables the program to accomplish more with a single instruction execution. Larger page sizes mean fewer indirect memory references to get at data and, therefore, shorter running times.

Larger word sizes also permit the representation of larger table position displacements without resorting to multiple precision arithmetic in the computer programs. Certain advantages are also to be gained if the word size is a multiple of the 8-bit length of the ASCII and EIA character code, since handling part programs and management information functions requires a lot of textual data processing.

The critical aspects of the processing problem should be carefully examined by an experienced programmer, so that accurate estimates of the program size can be made. Several sample program units that are representative of the processing problems encountered in the system should be coded in the language of the candidate computers, so that core memory requirements and running times can be estimated. Areas to which particular attention should be paid are input/output operations, table-position control operations, control instructions for programmed loops, and the control structure within the program, that is, how the executive programs activate subprograms that perform separable processing functions. In the input/output area, one function that should be carefully examined is the processing that must follow an interrupt from one of the controlled machines. This is the amount of time required to identify the interrupting device and to output a prepared number. This function is exercised frequently in the control of stepping motor drives and its timing is critical when the motors are moving at or near their top rated speeds, since the external counters must be reloaded within the time between the lapse of one counter setting and the time for the next motor step.

The required amount of core storage is a function of the complexity of the computer programs and the data that must be stored and buffered in the core memory for processing. Additional core storage must be reserved for certain auxiliary functions not directly related to the actual processing task. For example, program loading and debugging routines are needed for the development of new programs and they must generally reside in core memory with the actual application programs.

Most computers come with minimum core sizes of 4096 words ($4K$), although some recently announced computers can have their core memories tailored to virtually any multiple of small powers of two, for example, 128 or 256 words. Unless the computer is handling only the direct control of a few machines, it is wise to count on core memory sizes of at least 8192 ($8K$) words. The extra core storage is very useful for testing programs during the development stages of the computer-aided manufacturing system, and it can be employed as buffer storage to reduce input/output accesses during final operation.

For direct machine control problems, the most frequently used item of data is the part program block. In the punched-tape form, this consists of a sequence number, table positions in either absolute or incremental forms, and various miscellaneous control codes for activating discrete machine functions. One of the first tasks that should be undertaken in estimating computer requirements is to decide on the philosophy under which this information is to be handled. For example, if the positioning information cannot be represented within the numerical limits of the computer word size, then some form of double precision representation is required. If the dynamic range of the interface control units is exceeded by the maximum part program block position command, then the positioning command has to be broken down into a number of steps that add up algebraically to the original part program command. This breakdown can be incorporated into the data as it is stored on the random access memory, or the data can be broken down dynamically as part of the control cycle within the computer program. In the first case, more core and disk storage are required to hold the equivalent of a single block. In the latter case, more processing time is required during the control cycle to break up the position command. In estimating the amount of core storage required for these and other items of data, one should be sure to consider the problem of handling the data as well as attaining a minimum storage representation. One technique that is frequently used by the inexperienced is that of counting the number of bits required for the maximum size of each parameter, adding them up, and dividing by the word size of the computer. Maximum packing of the data is the storage representation that this technique implies and, while it achieves maximum data density, it fails to account for the fact that the computer programs must "unpack" the data before use is possible. The result is a great increase in the running time of the programs and additional instructions to handle the unpacking functions. It also contributes to the likelihood of errors during the development of the programs.

For systems handling management information functions, the core requirements are more severe, since the core has to provide room for additional programs as well as buffers for their data. The direct control and management information programs compete naturally for core and processor time resources. Executive programs have to be included to coordinate the distribution of these resources among the two classes of programs.

If the access to part programs and the timing constraints on this part of the operation permit (as generally they do), then some of the processing programs can be retained on the random access memory and called into core memory for running only when they are actually needed. A certain section of the core memory might be reserved for these programs and the calling of these programs can be managed by reasonably straightforward executive routines. This approach is called "overlaying" and it represents an excellent way to gain greater utilization of limited core memory resources.

After core storage requirements have been estimated, they may be used to measure the competitive cost effectiveness of computers that are competing for positions in the computer-aided manufacturing system. However, one should always temper these estimates with healthy safety factors. Generally, at least 20 percent should be applied, although factors of 40 percent are frequently used by companies with experience in these areas. It is quite difficult to estimate accurately all of the requirements during the early stages of system planning, and the functional requirements of the system occasionally undergo some evolution even after the computer configuration has been selected. The confidence that one has in the ability of the selected computer hardware configuration to meet adequately the system requirements is directly related to the precision with which the system functional requirements are stated and the degree to which the original statement of requirements matches the final statement of requirements.

The principal considerations in the area of peripheral devices and the input/output structure of the computer are the response speed of the computer to interrupts, the input/output transfer rates, and the capacity of the peripheral devices, particularly the device used to hold the part programs during manufacturing operation.

The candidate computer should be equipped with a priority interrupt structure such as the one we described in Chapter 4. The interrupt structure finds its primary employment in the machine control processing area and in the retrieval of part program blocks from the random access memory. One of the main problems is the distribution of devices over the interrupt levels and identification codes provided with the computer. Ideally, a separate interrupt line should be provided for each axis to be controlled. Since the number of interrupt lines provided with most minicomputers is generally some small power of two, such as four, eight, or sixteen, the system designer may find them a limiting factor if he wishes to control a large number of machines simultaneously. If there are more axes than available interrupt lines, special interrupt interface hardware may have to be designed to handle the problem. A typical device might occupy a single interrupt level and transmit a code to the computer that identifies the actual axis which caused the interrupt. The simplest code might consist of setting a bit in the word transferred to the computer. The position of the bit would correspond to the device axis and the computer program could determine the number by shifting the input word and testing for its high-order bit being set (an instruction frequently provided with minicomputers). The interrupt line to which the special device was attached would serve to identify the group of axes being tested, and the count of shifts until the high-order bit of the input word was set would serve to identify the axis within the group. An alternate approach would be to have the computer individually sense each device through an input/output test instruction until it found one whose status indicated that it was requesting an interrupt. Both of these techniques add time to the interrupt identification function, the latter more than the former; but they do provide the capability of conveniently controlling more manufacturing tools from a single computer. As we noted earlier, the amount of computer time consumed by this function should be carefully analyzed since it determines, and limits, the maximum number of axes that a single computer is capable of controlling.

The disk or drum memory used for the retention of part programs is another device that is critical to the successful performance of the system. The principal consideration here is the amount of storage that the disk provides versus the system requirements for on-line part programs. In a computer-aided manufacturing system that is largely devoted to the manufacture of relatively fixed product lines, the random access memory should provide enough memory capacity to handle the complete set of part programs for each part that constitutes the finished product. The representation of the part program block should be compressed as much as is reasonable, so as to conserve storage space. As we noted earlier, savings factors of five to nine can be had by omitting sequence numbers, storing positions in binary form, and packing miscellaneous functions into one or two words. The sequence numbers can always be reconstructed from the order in which the blocks are placed in the file. For printed circuit drills and component insertion machines, the part program size is not so critical since the programs are relatively small. For wiring programs, however, relatively large part programs are the rule.

Space for directories must also be included in the peripheral memory requirement estimates. These directories are needed so that the computer does not have to waste valuable time searching through large files for part programs. In order to avoid wasted space or special file maintenance programs, one should plan on using chaining techniques. This is especially true for job shop operations, where part programs are changed and replaced frequently.

If the file of active part programs is changed rather frequently, the system should include peripheral devices for the input of these programs at high speeds. If the programs are to be produced on the computer system using design automation programs, or if they are to be modified, some form of relatively high-speed output device should be provided. The least expensive and most versatile forms are high-speed punched-tape readers and high-speed tape punches. Tape loading and punching through the Teletypewriter at ten characters per second is much too slow for a relatively active system. As an alternative to punched tapes, the system designer might consider cassette magnetic tape drives. However, punched tape has the advantage of compatibility with other installations and also permits the use of part programs prepared for use with conventional numerical controllers.

For computer-aided manufacturing systems with management information functions to perform, the problem of estimating peripheral memory requirements is more complex. Here, the volume needed for inventory files, bill of materials and part explosion files, production records, vendor and customer files, and so on, competes with the manufacturing data base for peripheral memory resources, so that the structure and interrelationships of these files must be carefully planned. One of the first items of information needed for a proper estimate is the structural design of the files. This design is based on the functional requirements of the management information processing and the technical characteristics of the computer system to be employed. Therefore, several iterations of the design and estimation cycle may be required before a proper design with well-estimated resource requirements is reached.

The processing for management information functions also usually requires additional peripheral devices for the generation of reports and the input of the basic data that make up the files on which this processing is based. This is especially true for systems that handle inventory operations and part explosions for large and complex manufactured parts. Line printers, for example, are necessary if the management reports involve large amounts of data, such as complete lists of the current inventory. The original data on which these files are based are traditionally carried on punched cards that form the data base for existing business and management information processing systems. Punched cards are also a convenient medium for the preparation of the data and for its subsequent modification. Magnetic tape drives that are compatible with large- and medium-scale business data processing installations are also of value to the system. They provide both a means of communication between the computer-aided manufacturing system and the corporation's business data processing system as well as a means for the retention of back-up records, so that important files can be reconstructed in the event of a serious system malfunction that results in damage to the on-line files.

For very ambitious systems, timing, core memory, and peripheral memory constraints and costs can present severe problems to be overcome. The competition between management information files and part program files for peripheral storage can be resolved with high-speed random access disk and drum memories serving as buffer storage for data during its active processing. Moving head disks are less expensive than fixed head disks on a cost per word basis. Thus the moving head disks can be used for the bulk of the storage. When data, such as a part program, is needed for an operation, it can be transferred to the smaller head-per-track disk. Timing problems, especially in the area of direct machine control, can be alleviated by employing multiprocessor or distributed computer systems where several minicomputers handle the direct control of machines under the supervision of a larger central computer which is responsible for management information processing.

The solution of the computer requirements question can be obtained only through extensive analysis based on well-defined requirements for the functional characteristics of the system, but there is feedback between the two. The functional characteristics may sometimes exert modifying influences on the requirements. The initial approach to the design of the system must not be taken lightly—the first step is the most important.

Program Design

The computer software makes up the heart of the system. It serves both as a coordinator of the system activities and as an economical and flexible processor for most of the functions that must be performed by the manufacturing system. To handle this function properly, the software must be carefully designed and based on a complete set of functional specifications. The structure of the software system and the internal rules under which it operates are the key to efficient system operation and minimum cost for development and expansion of the system.

The contrast between the computer-aided manufacturing system and the collection of conventional numerically controlled tools is an important one to keep in mind. Rather than having one autonomous controller for each manufacturing tool, we now have a number of computer programs operating in one or more computers. Each program performs a function that is common to the operation of each machine, and it performs this function for each machine under the control of the computer in which the program is running. Thus the programs are shared among the machines. The coordination and synchronization of these programs are not simple matters and the development of software for a complete computer-aided manufacturing system is generally the most costly single item in the complete system. For this reason, a great deal of concentration should be given to the software design; it should be monitored very carefully during the development of the system.

Modern real-time program design techniques call for the operation of functionally oriented routines and subroutines under the control of executive and operating system programs. A routine is a sequence of computer instructions that carry out the processing required to accomplish a well-defined function or related group of functions. The name "subroutine" is generally reserved for a routine that exists outside the main line of processing and can be called at any point in the progression of the program to execute a certain function. A subroutine is called through a form of branch or transfer instruction that leaves a "trail" in the form of a core memory address to which the subroutine returns after it has completed its processing. As we discussed in Chapter 4, this return address is generally the core memory address of the instruction that follows the calling instruction, so that the subroutine processing is performed between these instructions.

Subroutines are written to perform certain frequently needed functions, so that calls to the subroutine can be substituted for the detailed set of computer instructions which would perform the same function. The primary intention here is to save the time and trouble of writing the same sequences of instructions every time the function is required by the processing, and to conserve core memory by having one copy of the routine instead of many. It also has the effect of providing useful "black boxes" that can be interconnected, through the services of an executive program, to formulate a complete system of software. The analogy between back panel wiring interconnecting printed circuit boards and executive programs interconnecting subroutines is a very good one for visualizing this concept. The principal difference is that a single subroutine can serve its function many times at different points in the logical flow of events, while a separate printed circuit card is usually needed each time its function appears in the logical flow of events.

The concept of functionally oriented subroutines under the control of the executive routines has the modularity that is very desirable. This characteristic makes testing of the system easier, since individual routines can be tested separately before they are incorporated into the larger system, where the isolation of programming errors may be more difficult. Modification of the system for future improvements is also made easier since the functions are isolated to particular routines, and changes made to these routines are insulated from the other routines in the system. Beyond these, certain managerial advantages accrue to the development phases of the project in that the tasks of developing individual routines can be more easily assigned to the programmers working on the project. Each programmer can be given specifications for his assigned subroutines, which he can write and test with a minimum of communication with other programmers. Of course, these specifications must be developed in great detail before the coding and testing is begun; the development of these specifications requires consideration of the relationships between the functions performed by the subroutines that make up the complete system.

The first step in setting up the structure of the computer programs is determining the most natural and efficient flow of data through the system. Functional subroutines are then identified with the operations required to effect this information flow. Thus the subroutines are arranged around the information flow. One possible data flow for the control of machines is illustrated in Figure 5.10. We find the part program placed in the data base file, from which part program blocks are retrieved sequentially. The position commands from the part program blocks are broken down into positioning steps that match the dynamic range of the interface control device; they are sent to the interface control device as separate positioning commands. The interface control device translates these commands into motion of the machine under its control. When the positioning step has been completed, a step complete signal is used to select the next positioning step. If the total motion called for by the block has been completed, the step complete signal causes the output of commands for the machine discrete functions. Thus we can identify the following functions to be performed:

1. Loading of the data base file
2. Transfer of blocks from the data base file to the computer

3. Breaking up of the part program block into positioning steps
4. Transmission of positioning commands to the interface control
5. Recognition of the step complete signal
6. Output of machine discrete function commands

We can design subroutines to handle each of these functions. They would be coordinated by an executive program responsible for the general functional area of machine control. We should also note that an element of

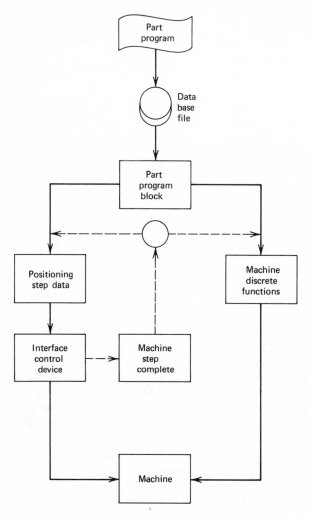

Figure 5.10. Typical data flow for machine control.

coordination and synchronization is also induced by events external to the computer, namely, the completion of a step by the machine. The details of machine discrete operation can be expected to vary from one machine to another and it is here that the modular characteristic of subroutines can be used to good advantage. The portion of the executive responsible for this part of the operation can be structured to call on different subroutines depending on which machine is currently being dealt with. Thus a separate subroutine can be designed to handle the discrete outputs for each type of machine. When a new machine is installed in the system, a subroutine for the discrete control of that machine can than be plugged into the software system with very little trouble.

There are four principal classes of programs in this kind of software system. The *executives* are responsible for the coordination and synchronization of the other programs in the system. The *interrupt response programs* are responsible for the recognition of computer interrupts and the execution of whatever responses are immediately required by the interrupting event. The *input/output handlers* are responsible for the initiation of data transfer to and from computer peripheral devices, and error checking and recovery procedures. The *processing programs* are the routines and subroutines that handle processing for all of the other functions, especially those peculiar to the system's requirements. This class includes routines that handle machine discrete functions, management information reports, file preparation, and design automation functions such as creation and modification of part programs.

These programs communicate with one another through subroutine calls, data values, and programmed switches retained in core memory locations. The placement of these locations is quite important, especially in small computers with limited core memory addressing capabilities. Insofar as possible, they should be located in memory pages to which the programs have the most efficient access. This is generally the base page, as described in Chapter 4. This page is the main page that can be directly addressed from any location in the core memory. Locations in other pages are usually restricted with regard to direct access. The only points from which direct access is possible are locations in the same page. Thus the general principle is to place commonly used parameters in the base page and to have subroutines keep locally used parameters in the vicinity of the subroutine.

The assembly programs, which are supplied by the computer vendor, generally provide features that facilitate communication between separate routines. The most typical feature is a pseudo-operation which allows the programmer to identify a parameter as being defined "external" to his own program. This parameter is actually defined in some other program and the name and location of the parameter are provided to other routines through another pseudo-operation in the program that creates a special assembly output for that location. The locations are identified in the programs by symbolic names and this assembler output consists of the symbolic name of the location in a separate and specially marked block of output. The loading program then takes over responsibility for tieing the whole thing together. The loader maintains a table of these names in core memory, building this table as it loads segments of the programs into core memory. When it comes to a block identifying one of these names, it enters the name in the table along with the location assigned to the name. When the loader comes to a block that requests access to this symbolically named location, it uses the location address stored in the table to formulate the machine language instruction. If the disposition of the location and the instruction in the core memory is such that direct access is not possible, the loader formulates the instruction as an indirect access and places an indirect reference word in a core location that can be directly accessed by the instruction. In most cases these indirect reference words are placed either at the end of the base page or the end of the page containing the instruction. Subsequent indirect reference words are stored working backwards from the end of the page. If there is competition between the instructions and these indirect reference words (i.e., they end up trying to occupy the same word of core memory), then the program cannot be properly loaded and the loader generally provides some kind of printout to inform the programmer of this fact. In this case, the program must be revised to prevent the conflict. In estimating the size of the program and the consequent core memory requirements it is necessary to consider the access problem and the amount of locations required for internal tables, pointers, and programmed switches (called "flags" in programming parlance). These considerations also affect the running time of the program since each indirect reference adds one memory cycle to the running time of the programs.

The concept of foreground/background operation is also quite important to the real-time control system. The terms correspond to the priority given a routine for central processor time. The interrupt programs, for example, have the highest priority in that they are initiated immediately by external events. Programs that must accomplish their processing functions under timing constraints are given the next highest priority. In general, programs whose priorities are assigned with regard to timing constraints are termed "foreground" programs. All other programs operate in background.

This approach to the time structuring of computer programs is generally employed in computer-aided manufacturing systems. The interrupt response programs, which handle the transfer of positioning commands to the interface control devices, operate in the foreground. Other interrupt response programs are responsible for detecting the completion of input/output operations, so that input/output transfers can take place in parallel with other processing; however, their work is generally restricted to the setting of flags that indicate the completion of the operation. If a time base is included in the system, an interrupt program is required to handle this function. This is mechanized by some form of pulse generating logic that periodically interrupts the computer, for example, once per second. The interrupt response program that handles this "clock" uses the interrupt event to count out seconds, minutes, and hours, so that relatively accurate clock time can be associated with certain system events. For example, this time can be used to measure output of machines over prescribed periods, initiate special end of work shift procedures, and so on.

Programs that retrieve part program blocks break them down into steps that match the dynamic range of the interface control devices and handle the machine discrete function control also operated in the foreground; but the timing constraints on these programs are not as severe as those placed on interrupt response programs. The execution of these routines is generally initiated by interrupt response programs when the interrupt response program determines that some function has been completed. For example, the interrupt program may have detected that the machine has just completed its positioning cycle. The next step would be the execution of the machine discrete functions and the interrupt program would set a programmed switch to indicate that the latest positioning cycle had been completed. The executive responsible for machine control would then recognize that the machine discrete functions were the next operation to be done by that machine and call the subroutine responsible for handling that function. There might be several machines waiting for the execution of their discrete cycles and the executive would determine the order in which these were performed. These functions should be completed as soon as possible, so that the manufacturing process can be kept moving at the fastest rate. For this reason, these functions are also placed in the foreground processing area.

Programs that do not have strong timing constraints are placed in the background processing area. When there are no more foreground processing tasks to be performed and the computer is in the situation of waiting for further interrupts from the controlled machines, the background programs can begin or continue their processing. This background processing can continue until the next computer interrupt occurs. Once the processing required by the interrupt or its consequences has been completed, the background programs resume. The resumption of background processing is another function of the executive programs which must recognize that the foreground programs have no more immediate work to do. Since both foreground and background programs can be interrupted, sorting out the proper place at which processing is to resume is no easy task. Eventually, of course, the originally interrupted program must be allowed to resume, but many other programs may have to operate in the meantime and many other interrupts can occur during that time. The programs assigned to the background processing area are most frequently management information processing programs and design automation programs.

The proper utilization of the executive programs and the services of operating utility programs is important for the successful operation of the system software. Beyond the executive programs, several operating system programs must be provided that perform useful functions for the other programs in the system. Certain input/output device handlers fall into this category. For example, printer output programs that print management information reports, Teletypewriter control programs that handle communication with the operators of the system, and disk input/output programs should be developed and be capable of servicing the requirements of the other routines in the system. Since several programs want to communicate with these devices during the execution of essentially independent processing functions, some form of competition resolution has to be incorporated into the design of these utility programs. Requests for disk data, for example, have to be sorted out according to the priority of the calling programs. Machine control programs calling for part program blocks from the disk should have a higher priority than background programs seeking records from an inventory file.

Another service that may be required of the operating system programs is the management of program "overlays." The technique of overlaying has been mentioned as a means of dealing with the constraints of small core memories. By retaining certain programs on the disk memory and calling them into core memory only when they are actually required, the system can effectively increase the number of functions that it can perform beyond those naturally permitted by the actual size of the core memory. A certain small section of the core memory can be reserved for these programs. When these programs are called for, either through an operator command or as supporting functions for other processing, overlay control subroutines can be called that (using directories) locate the requested program on the disk, cause it to be loaded into the reserved area of core memory, and execute it. This must necessarily be limited to background programs, since the time taken by the disk-to-core transfers is relatively long and does not generally find itself compatible with the requirements of processing in which the time constraints are strong.

The file handling techniques that are employed in the software system are also important. In general, they should be designed to fit in with the priority concept and the general control problem. For example, no program that handles processing functions should execute its own input/output operations. It should be forced to employ the services of the operating system programs. In this way, usurping of higher priorities by low-priority programs is prevented. All input/output processes should employ the interrupt structure of the computer as much as possible, so that processing can take place in parallel with the transfer of data to and from peripheral devices. If a program is to handle a number of data file records sequentially, then it should retrieve as many of these records as possible with a single transfer operation. A good example is the machine control program which is responsible for the retrieval of part program blocks from the data base file. By retrieving several blocks with a single request, the total amount of time to retrieve the complete part program is reduced and more time is left over for other processing. This technique of buffering data may also be employed by the programs that output to peripheral devices. The principal constraint on the application of this technique is the amount of core memory that can be reserved for the storage of this information.

As we stated at the beginning of this section, the computer software makes up the heart of the computer-aided manufacturing system. It is therefore to be expected that the time spent in this area of design is quite important. All parties to the design should participate in the structuring of the software, so that all aspects of the system requirements are properly met.

DESIGN EXAMPLE

We now discuss a design example that illustrates and summarizes some of the principles and concepts of computer-aided manufacturing system design. We cover the general design of a complete system for the control of electronic manufacturing machines of the numerically controlled variety.

The general organization of the system is illustrated in Figure 5.11. It is similar to the approach shown in Figure 5.9, especially in the modularity of the computer interface logic. The central control element in the system is a minicomputer that, in its initial configuration, has 4096 words of core memory. Two Teletypewriters are provided. One serves as the control console for the system and is positioned near the manufacturing area. The other reports manufacturing information to the supervisor or manager and is located in his office. A head-per-track disk memory is provided for the retention of part programs and a high-speed punched-tape reader is provided for loading the disk memory with these part programs.

The computer interface rack contains device selection gating logic, com-

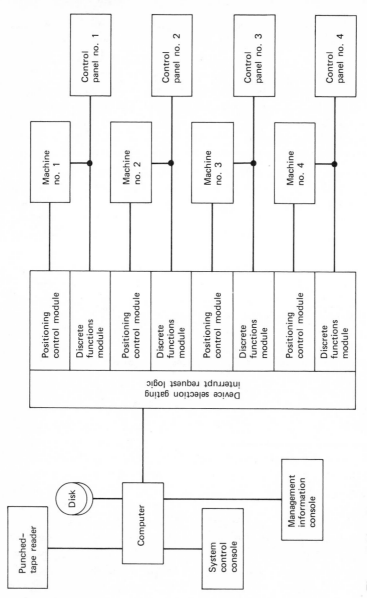

Figure 5.11. General system organization.

172

puter interrupt request logic, control logic for machine axis actuators, and logic for the transfer of discrete function signals to and from the computer. The interface rack contains one module for the positioning control of each machine and one module for the discrete functions associated with each machine. Power supplies, gating logic, and interrupt request logic are shared among the machines connected to the rack. The indexers for each axis are positioned near the machines. Stepping motors are used as actuating elements with open-loop control.

The position control logic is identical for each machine that utilizes two-axis positioning. This logic controls wire termination machines, printed circuit board drills, and most component insertion machines. Machines having more than two axes can be controlled with combinations of these logic modules which are coordinated by special software. Applications requiring more than two axes include automatic drafting machines for printed circuit masks and some component insertion machines that permit rotation of the board for nonorthogonal component placement and variation of component center distances.

The discrete function logic modules are different for each type of machine. These modules are principally collections of flip-flops and are used either as buffered drivers for relays in the machines or as buffered sense elements for limit switches and operator controls on the machines. For machines that require display functions or manual positioning inputs, these functions can also be handled through discrete function logic modules.

Figure 5.12 illustrates the position control logic for one machine. At the start of a positioning cycle, the computer loads the N-stage counter with the number of steps to be taken. If the motor is being brought up to speed, this count may be less than the full range of the counter. This is also the case for the deceleration portion of the positioning cycle, as the motor approaches the desired end point. The number loaded into the counter is a function of three variables: the present stepping rate of the motor, the distance that remains to be covered, and the acceleration/deceleration capabilities of the motor. We discuss the determination of these counter load values shortly.

The stepping rate is determined by a computer-controlled clock. There are a number of ways that this clock signal can be obtained. We have shown a rate multiplier which multiplies the signal (obtained from a single clock common to all devices) by a binary fraction which has three bits. A voltage-controlled oscillator could also be employed and might be controlled by a 3 or 4-bit output from the computer, converted to an analog voltage. The choice here is determined by the relative costs of the devices. The functional characteristics are identical in both cases. The design that we are discussing uses motors whose acceleration curves can be reasonably

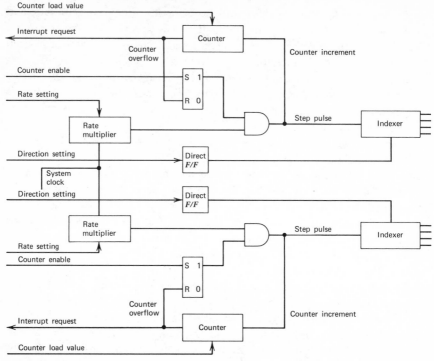

Figure 5.12. Position control logic.

approximated with seven points placed on a straight line from the start/stop rate to the maximum rate. For some motors, better approximations to the acceleration curves may be required and, therefore, more selection of rates must be provided. This means that more bits must be provided in the binary fraction that controls the rate multiplier.

The computer loads the counter with the 2's complement negative of the number of steps. The operation of loading the counter enables the gate on the output of the rate multiplier. The counter for each axis is incremented at the rate selected by the program output to the rate selection logic. The incrementing pulse train is also used as an input to the motor indexers. The motor direction is specified by a discrete level direction setting that is established by a computer output at the start of the positioning cycle and left unchanged until the cycle has been completed. When the counter increments to zero and overflows, the signal to the indexer is stopped and an interrupt request is generated from the logic assembly. The computer then reloads the counter with the next position count and outputs the next rate setting. Since one counter can overflow before the other and may be controlling a motor operating at a rate greater than the start/stop rate, it is necessary to recognize the counter overflows for each axis independently.

The number of stages that the counter should have is a function of the computer time allocated to the direct control task. Generally, the more stages in the counter, the less total time that the computer has to devote to the direct control task. The time allocation aspect of this problem was analyzed in the section on Computer Responsibility. Limiting factors on counter stages are cost and the effect that the stages have on the acceleration/deceleration cycle. To analyze this problem further, let us set up some parameters. Let

n = number of bits in the rate control (linear control is assumed),
C = rate of signal input to the rate control logic (driving signal to rate multiplier),
R_o = start/stop rate of motor,
R_T = top rate of motor,
R = current stepping rate of the motor,
k = output word to the rate control logic.

The interpretation of the rate control logic is that it serves to multiply the basic rate C by a fraction of the form $k/2^n$. This interpretation can be taken whether the rate control logic consists of a rate multiplier or a voltage-controlled oscillator. Thus the current stepping rate is given by

$$R = \frac{kC}{2^n}. \tag{5.3}$$

We attempt to approximate the acceleration curve of the stepping motor by a series of points along the curve, one point for each value of the output word to the rate control logic. Ideally, the lowest value corresponds to the start/stop rate R_o and the highest value corresponds to the top rate R_T. Generally, the acceleration curve is nearly linear. Thus the individual values of rate control serve to divide the total acceleration time into $2^n - 2$ intervals of equal length. This is illustrated in Figure 5.13.

The requirement that the lowest rate control value correspond to the start/stop rate is given by

$$R_o = \frac{C}{2^n}. \tag{5.4}$$

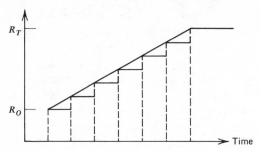

Figure 5.13. Ideal relationship between acceleration and approximating curves.

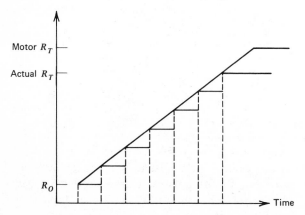

Figure 5.14. Actual relationship between acceleration and approximating curves.

The requirement that the highest value correspond to the maximum rate is given by

$$R_T = \frac{(2^n - 1)C}{2^n}. \tag{5.5}$$

Together, these imply that the start/stop rate and the top rate be related by the following equation:

$$R_T = R_o(2^n - 1). \tag{5.6}$$

The top rate cannot always be expected to be a multiple of the start/stop rate, so we cannot always hope to achieve an exact match at both ends of the acceleration curve. Since the start/stop rate is the more critical of the two requirements, we satisfy this one and settle for an actual top rate that is slightly less than the rated top speed of the motor. This is illustrated in Figure 5.14. For example, with a 3-bit rate control word

$(n = 3)$ and a motor with a rated start/stop speed of 200 steps per second, we have the following selection of speeds available:

k	Rate (steps per second)
1	200
2	400
3	600
4	800
5	1000
6	1200
7	1400

If the counter in the position control logic has m stages, then the time between interrupts at stepping rate R is given by

$$I = \frac{2^m - 1}{R}. \tag{5.7}$$

For the example given in the table above, at the top speed of 1400 steps per second, we have slightly over 714 microseconds per step. Thus for a four-stage counter, there would be about 10.7 milliseconds between computer interrupts. For an eight-stage counter, we get approximately 182 milliseconds between interrupts.

As we stated earlier, the principal motivation behind external position control counters is to provide more central processor time for the performance of the other functions required of the computer. By servicing machine axes less frequently, less processing time over a particular period would be devoted to the direct machine control function. With the four-stage position counter, an axis operating at top speed requires service from the computer approximately ninety-nine times per second. During the acceleration cycle of an axis, the time between interrupts may vary. The reason for this is the attempt to maintain the acceleration curve that we have described. If the total time from start to the attainment of top speed is D seconds, then the acceleration period is divided into $D/(2^n - 2)$ second intervals. During each of these intervals, the driving rate to the motor is held constant. At the end of each interval, the rate is increased to the next rate in the approximating curve. The interval time length is controlled by loading the counter with the number of steps that the motor takes at the given rate during the required interval. Thus if the rate is R, the value loaded into the counter would be $DR/(2^n - 2)$. If the time to reach full speed is 60 milliseconds, with a 3-bit rate control word we would have six intervals of 10 milliseconds each.

Since the number of machines undergoing acceleration or deceleration and the number of machines with axes operating at top speed is a function of operator responses and the characteristics of the part program being executed, we can consider the demands with regard to computer attention to be somewhat random. In general, a reasonable approach to the problem is to examine the characteristics of the physical motions required during the manufacturing operation and thereby arrive at some kind of typical motion command. For example, holes drilled in a typical printed circuit board might be on 0.5-inch centers. Thus 0.5-inch would be the typical motion for this kind of machine. For semiautomatic wiring, the typical wire might be 2 inches long and this number could be reasonably adopted as the typical motion for a semiautomatic wire termination machine. The total amount of time required for the machine to carry out one of these positioning cycles is determined by the typical positioning speed. Given these numbers and the mix of machines in the system, one can then determine the number of times the computer has to service a machine to accomplish one of these motions. At this point, one should estimate the amount of time required by the computer to service one of the machine interrupts. This should be done by drawing out a rough flowchart of the operations that the computer must perform to prepare and output the next counter setting. From this flowchart, a programmer can code a sample program that carries out these operations, and the running time of the program can then be estimated, using the memory cycle time of the computer. The product of the running time of this interrupt response program and the number of times that a typical machine requires service during the execution of a typical part program block give the total amount of time that the computer must devote to servicing the machine during this time. The total amount of servicing time divided by the positioning time of the machine yields the fraction of that time devoted to the control of the machine. This fraction is an indication of the loading on the computer. Dividing this control time into the duration of the typical positioning cycle yields the number of machines that can be controlled without any machine having to wait (on the average) for servicing in starting up the next part program block. Ideally, some fraction of this time should be reserved for functions other than direct control, the actual amount being determined by the amount of processing required for these functions.

The other principal consideration is the hard limit imposed by the open-loop control scheme and the fact that the pulse train to the stepping motors must be fairly regular. In general, this means that the computer must reload the position counter in the interface device before the time for the next index pulse comes up or during that time. If the computer were to delay for one or more pulse periods before reloading the counter, the motor might very well take an extra step that would go uncounted. An occasional error of this type might cause no difficulty; however, if they are due to continued overloading of the computer, they are likely to accumulate into considerable errors over the duration of the operation. The best protection against this is to ensure that there is adequate computer time during fully loaded conditions to service all machine axes within one top rate interpulse period. In the example that we have been discussing, we have approximately 714 microseconds for the interpulse period. This time divided by the interrupt response time estimated from the preliminary coding of the program yields the number of axes that can be controlled simultaneously. Two choices are then available to the designer. He may either limit the number of machines that will be controlled by the system or the computer program can be set up to limit the number of axes in simultaneous motion. With the latter technique, a machine about to begin the positioning cycle would have to wait until there were fewer than the maximum number of allowable machines operating before beginning its cycle. In practice, this waiting time is rarely noticed, since times during which all machines are operating simultaneously are infrequent.

Figure 5.15 illustrates the general organization of the computer programs. The executive routine handles control of all of the programs except the position control routine, which is interrupt driven. It communicates with the system operator via the system control console, which is a Teletypewriter. The operator uses this console to assign part programs to various machines in the system and otherwise control its operation.

The position control routine is initiated by the interrupts from the position control devices and is responsible for obtaining the next loading value for the counter, outputting it to the position control device, and recognizing when a particular machine has reached the commanded position and is ready for its discrete function outputs. The machine sequence routine is responsible for handling the machine discrete output and input functions and for the initiation of the next position cycle. When the buffered set of part program blocks for a particular machine has been used up, the machine sequence routine calls on the part program block retriever to get the next set of part program blocks. The part program block retriever reads these blocks in from the disk memory, where the part programs are stored. The part program file handler is responsible for reading new programs from the punched-tape reader, recording them on the disk memory, erasing old programs, and maintaining current directories of the part program file. The management information routine is executed periodically and monitors the progress of the machines under the control of the computer by such techniques as counting the number of times that a particular part program is executed. Counts and indicators maintained by the other routines contribute to the data that the management information

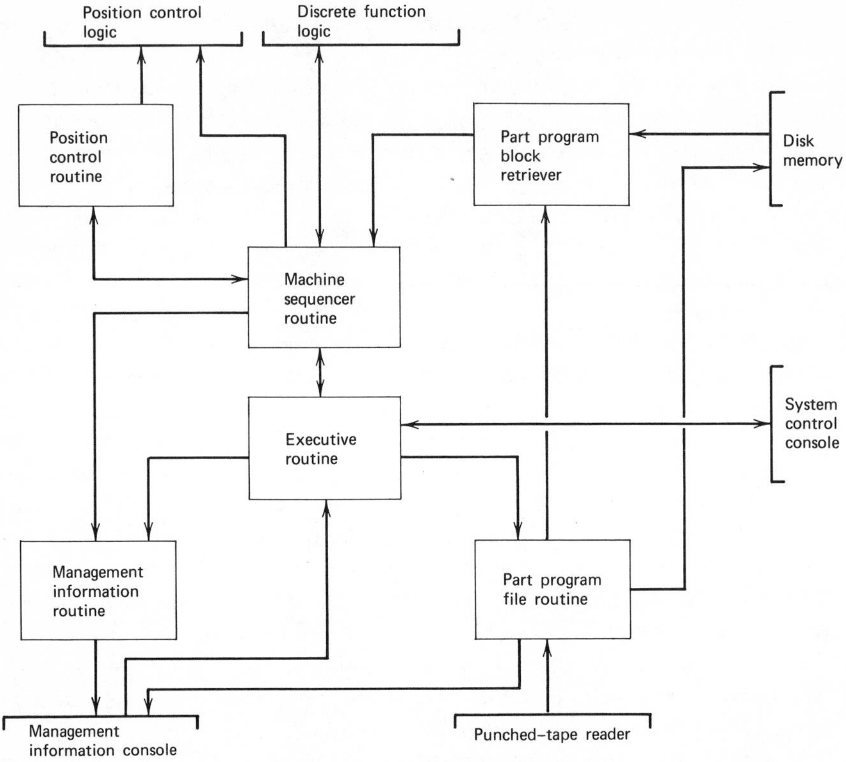

Figure 5.15. General organization of computer programs.

routine works from. From this data, the routine can prepare reports on the total number of parts produced by each machine in the system during the current reporting period, the total output of the system, and the current status of any particular piece being produced (such as the number of wires in place and the number of wires remaining in a wiring job). This routine can also report on the availability of machines, the reasons for an idle state, and so forth.

SUMMARY

In this chapter we have discussed some of the technical principles that underlie the design of a computer-aided manufacturing system. A great deal of detailed analysis must go into the final and complete design of such a system; we have only touched the surface here. The economic and technical advantages of various approaches can be expected to change with the development of new component and manufacturing technologies and their effects on the cost and availability of the materials and devices from which such a system is fabricated. One of the principal factors is the continuing price reductions for minicomputer systems. At the present time, for example, it is almost the case that one might use entirely separate minicomputers in the place of the interface control devices that we have described, with the central computer (still conceivably a minicomputer) performing the management and part program handling functions. In such a configuration, it would be reasonable to load heavily the control minicomputer, even to the extent of having no counting devices in the interface and using the direct generation of bit patterns or pulses to indexers driving stepping motors.

In the next chapter we discuss the integration of the complete manufacturing process into a unified whole under the control of computers, the advantages of such a system, and the means of justifying it.

Chapter Six

Integration of the Manufacturing Process

The purpose of this chapter is to discuss the way in which a computer-aided manufacturing system fits into the manufacturing operation and the remainder of the corporation as a fully integrated element. By observing the effect that the CAM system may have on the manufacturing operation and corporation, we can judge the relative merits of adopting (or not adopting) various kinds of CAM systems. Such systems may have varied effects on the manufacturing operation and the warehousing, purchasing, and distribution facilities that support it. The ways in which these operate, and the condition of the marketplace, help to determine the optimum way for the corporation to allocate its resources with a view to maximizing the return on its investment. One possible allocation is an investment in computer aids to the manufacturing operation and its supportive facilities.

Maximizing return on investment is the principal objective of the manager who is doing his job properly. If he is interested in computer-aided manufacturing systems for the purpose of "empire building," or because they represent attractive state-of-the-art techniques, then he is motivated for the wrong reasons.

The principal function of the computer is handling information. The computer handles this information according to rules represented by the programs that operate in the computer. The function of information handling is present even when the computer is controlling machinery—although it is not so obvious then as in the analysis of operational data and the preparation of reports for the management of the corporation. That the principal concern during the integration of computer aids into the corporate and manufacturing organizations is the handling of information should not be very surprising. Throughout this chapter we are concerned with the ways in which information is handled.

We begin by considering the objectives and responsibilities of the manager who is associated with a manufacturing operation. We then discuss the integration of the CAM system into the present manufacturing system and into the overall corporate structure. From the principle of maximizing return on investment come motivations to enhance the manager's effective control over the manufacturing operation and to enhance its productivity. The attainment of these goals represents reasons for the installation of a computer-aided manufacturing system. We discuss some aspects of enhancing productivity and effective control and finish the chapter with a short discussion on the problem of system acceptance.

MANAGEMENT OBJECTIVES AND RESPONSIBILITIES

Generally, the manager and staff of an electronics manufacturing organization are retained at rather healthy salaries for the purpose of fulfilling certain functions pertinent to the objectives of the corporation. The principal objective of the corporation is maximizing the return on its stockholders' investment. It is important, therefore, for the manager of the manufacturing operation to be very aware of the effects of his decisions on this aspect of the corporation. The function of manufacturing management is to control the operation of that part of the corporation that actually fabricates whatever it may be that the corporation purveys in the marketplace in order to realize its income. The cost of the manufacturing operation affects directly the income realized from these products. The less these products cost to make, the better the general situation of the corporation, and the closer the manager is to properly fulfilling his function. Therefore, control over the cost of the operation is one of the first objectives of the manager. From this basic objective other objectives can be derived.

The manufacturing operation does not stand strictly alone in the structure of the corporation. There are many tendrils connecting the manufacturing organization to other elements of the corporation. There are many channels through which materials flow—raw materials entering the manufacturing process to be converted into finished goods which must then be distributed in the marketplace. The manufacturing manager must therefore be aware of events taking place in other sections of the corporation. How these sections operate, and the degree to which they satisfy their functions in the corporation, influence the degree to which the manufacturing operation successfully performs its function.

These operations coordinate their activities through various channels of information. The sales organization provides feedback to the corporate system from the marketplace, where the general acceptability and success of the product are measured in terms of sales placed. The quality of the product is also measured through contact with customers and this measure must have an effect on the operation of the manufacturing facility, for example, through detection of a need to improve quality control practices in the face of malfunctions in the field. The general management of the capital resources of the corporation and its tax situation depend to a great extent on the volume of output from the manufacturing operation and the amount of time that elapses between fabrication and actual delivery to customers. The manufacturing operation is fed by various sources of raw and semifinished materials and the control of these sources is the responsibility of other segments of the corporation, namely, the purchasing and procurement operation. These operations control their own volume of material handling to match the demands of the manufacturing operation. Thus the accurate measurement and forecasting of these demands is another important concern of the manufacturing manager.

The manager is always attempting to control a collection of people and machines that constitute a system. This control is based on information that the system provides about itself and information that is provided about the system from sources outside it. In the greatest number of cases, this information is provided by people, and the channels that the manager has available to him for the control of the system lead to people who are part of the system. For these reasons, the manager must concern himself with the relationship that he has with his employees and with the relationships they perceive, both with him and the system in which they work. If the manager can improve the channels of communication with his people, then he has every reason to believe that this will help him attain his principal goal of profitable operation.

A large body of tradition surrounds the organization of production facilities. There are many types of production facilities, with examples of most to be found in the electronics industry. Generally, there is a hierarchy within the corporate structure that includes the manufacturing operation. This hierarchy is composed of a high-level manager and a number of assistants, who are usually organized around the components that make up the product manufactured by the organization. Working under these assistants are various supervisors of one thing or another. At the bottom level of the hierarchy are workers, operators of machines, wiring people, assemblers, testers, and so on. At any given position within the organization, events take place that are stimulated by decisions. These decisions may have been made in the distant past. In this case, they are represented by established policies and procedures. They may also be made in the heat of some present crisis. Every detected event results in some kind of decision. This decision leads to further events—which lead to further decisions. We sometimes find situations where decisions are being made even in what may seem to be a managerial vacuum. Even continuing to do the same thing in the presence of no significant change in the state of the system may be considered a decision. This is going on in many well-managed operations where the system is sufficiently thought out, and provided with well-understood and viable procedures to handle almost any situation, that it virtually runs itself. If the correct reaction to a given set of circumstances is understood by as many people as possible, the system has a natural tendency to operate on its own. To get his own system into this kind of condition is also a legitimate goal of the manufacturing manager. One characteristic of such a self-propelling operation is the generally free and accurate flow of information among the people who make up the system. Given that the goals of the personnel agree with the goals of the organization, and given the existence of unambiguous information and policies on which to base decisions, many of these decisions can be made at levels close to the points where they are actually implemented.

One characteristic of the organization is that the heaviest flow of information is between points that are close together. The manager, for example, tends to have the most communication with his immediate staff. The worker tends to have the greatest amount of communication with his coworkers and immediate supervisor. When information originates at some point in the organizational structure, the rules generally require that it be passed from each level and point to the next, never directly between the source and the ultimate destination. Sometimes natural channeling effects tend to send the information to some points more quickly than to others, and occasionally conspire to prevent the arrival of the information at certain points in the organization. For each stage through which the information must pass, a time delay naturally occurs. If the information must travel over a long distance in the organization, this delay can accumulate considerably. Thus one more goal for the manager is to create an organization in which information flows more quickly to those who must make decisions based on this information. If decisions can also be made at a point close to the control points in the organization, then they can be put into effect more promptly and the general responsiveness of the organization is improved.

The manager must also be capable of evaluating the performance of his operation in an objective way. To do this, he must have access to data that measure this performance. The data must also be formulated in a way that helps the manager pinpoint reasons for any failure of the operation to live up to its reasonably anticipated goals, so that the manager can correct the problems before they adversely affect his principal goals of profitmaking and the relationship between his organization and the rest of the corporation. In most corporations, the collection of this data is the responsibility of sections of the company assigned to an accounting function. This data is collected periodically and formulated into reports that go to the management of the corporation. These reports outline the performance of the corporation over the reporting period. In many large corporations, these functions are carried out over extremely short periods—often on a virtually continuous basis. Thus the manager of manufacturing for a very large corporation may have his own accounting staff reporting strictly on the performance of the manufacturing operation. Most corporations, however, perform these accounting operations over longer periods, generally monthly. The length of the reporting period and the amount of data that must be handled usually mitigate against the production of timely reports. In many instances, the manager must wait more than a month to receive the accounting reports that delineate the performance of the operation for which he is responsible.

The manager's principal reason for wanting to know how his operation is performing is to enable him to make intelligent decisions regarding its day-to-day operation. Essentially, he is fulfilling the function of the control algorithm in a control system. The time lag inherent in the collection and presentation of the data on which he makes his decisions has several deleterious effects on the efficacy of these decisions. First, the delay increases the time during which an unsatisfactory condition can be allowed to remain in effect. Second, the delay forces the manager to make predictions about future events based on historical data of some antiquity. If the reporting cycle is one month and the report contains the discovery of a condition for which a sure solution is not known, then two months must necessarily pass between the first occurrence of the condition and the first opportunity the manager has to see if the solution that he attempted was in fact successful. This delay represents a substantial percentage of the total annual operating time and a serious condition can make the difference between a profit and a loss condition.

All these managerial objectives and responsibilities are concerned with information—the rate at which it flows, its accuracy, its timeliness, its clarity (the characteristic that the information lends itself to for prompt and accurate interpretation, so that effective decisions can be based on it). Insomuch as the computer is a machine for handling information, it is reasonable to expect that the computer-aided manufacturing system provides the most useful services to the manager of manufacturing in the area of handling information. If the system can provide the manager with better information, then the intelligent manager should be able to utilize this information to make more effective decisions regarding the operation of the facility over which he has charge. The computer system does not, however, make these decisions for the manager. That function remains in his hands. The computer can be programmed to give orders based on the information that it handles, but the rules represented by the programs are generated by the manager of the manufacturing operation. It is his responsibility to set up these rules and see that they are incorporated in the computer-aided manufacturing systems that are employed in the manufacturing facility.

There are other responsibilities that go with the role of manager of manufacturing and virtually every one has some relation to information. Most of the traditional methods for computer analysis of production and corporate data have been well covered by software and computer systems offered by the major manufacturers of business data processing computers. The processing of business data in the traditional sense has generally required the handling of large files representing the transactions of the corporation during the analysis period. For this reason, relatively large memory units have usually been required for these systems. Since the computers employed for the analysis of this data are also used for other functions within the corporation, a premium is generally placed on the speed with which these computing installations perform their functions. As a rule, the cost of one of these installations prohibited having more than one computer installation of comparable size in any single facility of the corporation. The result is the placement of the business data processing facility under the aegis of the corporate official principally responsible for the collection and analysis of business performance and accounting data—the corporate controller or the treasurer.

When we consider a smaller section of the corporation, we find that the scope of its business data processing requirements and the volume of information participating in this analysis are substantially smaller than those of the corporation as a whole. If the economics of the computer installation justify it, it might be entirely feasible to install a modest computer installation specifically for the purpose of serving the business data processing needs of only that section of the company. In fact, the advent of the minicomputer has made such an installation economically feasible for many small organizations within the corporation. Business data processing software is now available for several small computers that are intended to operate in stand-alone configurations, serving the business data processing requirements of small companies and departments within larger corporations. The data handled by these small computer installations are generally obtained from the daily operations of the section in which it resides and, in the case of the manufacturing facility, much of the data can be obtained directly from computer systems that are directing the operation of the manufacturing machinery. We have already noted in Chapter 5 how these computers generate data that describe their influence on inventory files and the relation between the engineering data describing the product and the raw and semifinished materials that make up the product. These direct control computer systems can also report on the progress of any job that they are currently controlling and therefore constitute the basis of a data collection facility for a production control system. Thus they are naturally related to the information systems that already exist in every manufacturing system and can serve as a primary source of information for the manufacturing manager.

Each manager already has some type of system that he is using for the purpose of collecting data from his operation. In the most primitive organizations they are "word of mouth" systems, where production control information is passed from foremen to supervisors in *ad hoc* or regular meetings whose function is the evaluation of progress on the work then going on in the facility. The manufacturing manager is continually involved in this exchange of information and it consumes a large amount of his time. It therefore behooves him to try to make this process more efficient. The more efficient the exchange of information becomes, the more time the manager has to ponder the decisions that he must make. It also provides supervisory personnel with more time to perform their supervisory functions of directing the actual completion of the work. Improved flow of information from the supervisors to the manufacturing personnel can also improve the efficiency of the operation in that it leaves the workers more time for actual production and reduces the amount of time that they spend taking direction.

To understand fully how the computer-aided manufacturing system can assist the manager in fulfilling the responsibilities and in attaining the objectives that we have outlined here, we must investigate some of the characteristics of the manufacturing organization and the total corporate organization from the viewpoint of integrating the CAM system into these organizations. We can then investigate the ways in which the CAM system is used to enhance the productivity of the operation and the manager's effective control of the organization. The final acceptance of the system as an integral part of the manufacturing organization and the corporation is a function of the skill with which this integration is accomplished and the degree to which the system enhances the profitability of the corporation.

INTEGRATION INTO THE PRESENT MANUFACTURING SYSTEM

In previous chapters, we discussed the computer-aided manufacturing system from the viewpoint of engineering, concentrating on the technical details of using a computer to control directly the machinery that is used in manufacturing. We discussed the nature of the information that the computer system must have in order to perform this function and the nature of the information that is generated as a natural by-product of the direct control function. We had a situation in which the computer used engineering data that described the process of manufacturing the part. The information generated as a natural by-product of the computer function had primarily to do with the effect that the manufacturing operation had on the inventory of material feeding the operation and on the inventory of finished goods. The CAM system was also capable of reporting on the progress of any job that it was currently controlling.

The information generated by the CAM system is of value to other parts of the organization as well as the management of the manufacturing operation. One use of this information is in measurement of the efficiency and production rate of the manufacturing operation. This information also enables other elements of the corporation to coordinate their operations with the manufacturing operation. To see how this information is used, we look at the general structure of the manufacturing operation and the elements of the corporation that provide support to the manufacturing operation.

Corporations engaged in the manufacture of electronic products, like other manufacturing operations, exhibit certain characteristics of organization that are typical of any group of men and machines banded together to carry out certain functions. One of the principal characteristics is the division of labor, familiar to most of us from the history of the Industrial Revolution. Certain subgroups within the organization are given unique tasks. Their duties are the performance of these tasks according to procedures and conventions established as part of the operating rules for the overall organization. The mission of the corporation as a whole is attained, theoretically, through the cooperative efforts of all individual operating groups. In the case of the manufacturing function, these groups must co-operate in a highly synchronized manner for the operation to run smoothly. Furthermore, within single groups dedicated to general tasks we frequently find subgroups dedicated to separable portions of that task, or to tasks of a parallel nature in which the general functions of the group are similar to those of other groups but are directed to the handling of different materials. In the context of the manufacturing facility, for example, we might find that several product lines are manufactured by the corporation. Each product line may have a separate manufacturing facility or production line. Furthermore, each product may be fabricated from a number of complex components—each qualifying for treatment as a separate manufacturing task within the overall manufacturing complex.

Since the corporation consists of separate groups operating cooperatively to attain the prescribed goals of the corporation, there must be some means of assuring this cooperation. There are a number of individual men and machines within each group working to achieve the aims of the group to which they belong. These individuals must also cooperate and so we find that a means of directing and controlling the operations of these individuals is needed. Direction and control are the functions of a management structure which is superimposed on the working group organization. This management structure has a hierarchical form in that each level of managers performs its functions of control and direction under the supervision

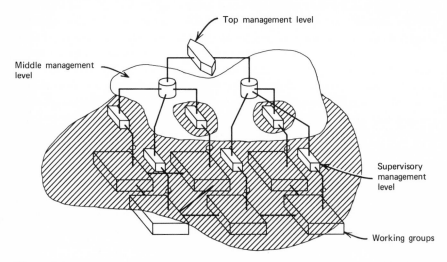

Figure 6.1. General corporate structure.

of still higher levels of managers. As we proceed further up the hierarchy, the number of members of each level becomes fewer. At the highest levels in the hierarchy, we find that only a few members are present and that they operate under the direction of a chief executive.

The general structure of the corporation that we have just described is illustrated in Figure 6.1. At the lower level, the working groups are shown with information and material passing among them. Above the working groups are several levels of management control, members of the first level each being responsible for a single working group. The function of this level of management is the direction of the activities of individual members of the working groups over which it has charge. This level is represented by the supervisory personnel in the organization, such as foremen. The next level of management is responsible for the control of several working groups, perhaps several production lines turning out the components of a single product. This level of management is generally termed "middle management" in trade journals and theoretical treatments of management science. The overall activities of the middle-management level, and therefore of the individual working groups, are coordinated by the top level in the management structure. Information flowing between working groups is of the tactical variety—information that is part of the normal everyday working environment of the groups. Information flowing from working groups to and among management levels is performance measurement and control information. This information is used in the establishment of policies and the making of decisions governing the operation of the working groups over longer periods of time. As long as the measurement information indicates no deviation from the planned course of corporate goal attainment, no significant action is required of the management level. When this information does indicate deviation from the planned sequence of events or quantitative results of operation, the management levels must decide and initiate corrective action to bring the organization back on its planned course. The channels over which these corrective actions are communicated carry control information to the working groups in the form of new orders or procedures. As we indicated earlier, the time cycle of measurement and corrective action may be quite long in some corporate organizations and the deviation of actual performance from desired performance can therefore accumulate a significant percentage of error. This error generally has some dollar value. It may manifest itself in excessive scrap cost, excessive inventory carrying costs, loss of sales due to sluggish performance, loss of customer goodwill due to quality control problem, and so on.

We call the structure represented in Figure 6.1 "stratification." We can see from this illustration that the corporation exhibits an effectively three-dimensional stratification. The stratification among the working groups alone is of a two-dimensional variety. Stacked above the two dimensions of the working group stratification is a third dimension of management strata. The working group strata can be termed "horizontal" and "vertical," according to their orientation. The terms horizontal and vertical have been used in a similar fashion in other contexts. For example, a horizontal market is one which covers segments of many different classes of customers. A manufacturer of computers might sell his product to the business data processing departments of corporations engaged in many different industries. Such a manufacturer would then find himself in a horizontal marketing situation. As a rule, the more general the application for a product, the more horizontal the nature of its market. A manufacturer selling to a vertical market would be one engaged in providing a total line of products that serves the needs of most phases of a particular industry.

In the context of stratified manufacturing operations and segments of the corporation that directly support them, the organization becomes somewhat more intricate, but the element of stratification still persists. Figure 6.2 illustrates a typical organization of a large manufacturing operation and its relation to various supporting groups from the corporation. Here, we have purchasing operations providing stimuli to vendors of raw materials. These raw materials are processed by receiving and inspecting operations and are placed in storerooms, where they await processing by the individual manufacturing operating. Each manufacturing operation produces a component part of the final product. These component parts are passed on to subsequent stages of the manufacturing operation for incorporation in subassemblies and final assembly into the finished product. Since the production rates of the separate manufacturing operations are, in general, not uniform, some form of buffer storage must be provided along the way for semifinished work (called work in progress). The finished product must then be stored to await sale through the auspices of the sales organization.

Vertical stratification is exhibited in the flow of materials through the sequential stages of the operation—from purchasing, through manufacturing, to sales and customers. The horizontal strata contain operations with functionally similar tasks. In general, we find that groups working within the same horizontal stratum are operating in parallel, using materials obtained from the next higher stratum and passing them on to the next lower stratum after they have been processed in some way. The most important characteristic of this stratification is that the principal channels of tactical data lie in the vertically oriented strata. Examples of such data are the following:

- Picking lists for the storerooms that feed the manufacturing facilities
- Receipts records that represent the transactions of entering newly received materials into the storerooms
- Packing lists that identify the contents of shipments received from vendors
- Purchase orders received from customers that stimulate retrieval cycles from the storerooms which feed the sales group
- Work progress reports that track the progress of particular jobs through the manufacturing and assembly cycles.

Compared to the flow of data along vertical channels, the flow of data along horizontal channels is relatively small. Data generated from points in the same horizontal stratum can be combined at lower strata in the coordination of the manufacture of a particular item, but very few data

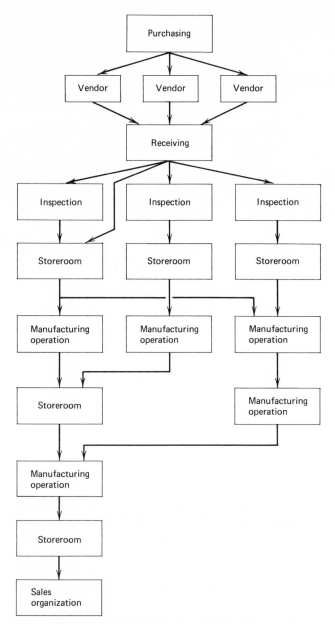

Figure 6.2. Typical manufacturing organization.

are actually transferred from one working group to other working groups on the same horizontal stratum.

The transactions represented by this vertically flowing tactical data form the basis for measuring the performance of individual working groups and the performance of the coordinated organization. Periodic summary analyses of these transactions are the principal content of information transferred from the working group plane up to the management strata.

Let us suppose that a computer-aided manufacturing system is installed in one of the working groups represented in Figure 6.2. What effect can we expect it to have on the operation of that group and on other working groups related to the group with the CAM system? If the system functions cover only the direct control of machinery (in the manner of numerically controlled machine tools) and do not involve the handling of tactical information, then the other working groups see little effect. The group in which the CAM system is installed sees benefits that normally accrue to improvements in the manufacturing rate and quality control that come from employing machines for this kind of work. It also sees some change in its capital worth, depreciation schedules, and other accounting burden factors. If the machines operating under direct computer control manufacture parts with fewer errors and reduced scrappage rates, then it will see an improvement in its productivity in those areas. However, inasmuch as the CAM system is not generating any new information regarding the work that it is doing, the general nature of the information flowing to other working groups should not be expected to change very much.

Certain second-order effects may be observed in working groups in the strata above and below the group with the CAM system. For example, an increased rate of manufacturing creates a heavier demand for parts from the storerooms that feed the manufacturing group. This increased demand may create some problems for the storeroom manager in keeping enough raw material on hand to feed the manufacturing operation. If the manufacturing group with the CAM system is being fed with semifinished work from other manufacturing groups, its increased manufacturing rate may similarly make demands on these groups to produce the semifinished goods at a greater rate. We noted that in some cases buffer storerooms were needed for the temporary retention of semifinished goods between manufacturing operations. It is conceivable that the improved production rate of a manufacturing group fed by semifinished materials may eliminate the need for such a buffer storeroom. We may also find that the improved quality of the product turned out by the system-equipped group may reduce the load on quality control groups responsible for the examination and testing of the output from the group.

Each of the effects that we have noted arises from changes in the rate and quality of the output from the working group equipped with the CAM system They have little direct relation to the aspect of information. If they had come to pass through the installation of new machinery that improved quality and manufacturing rates, but did not handle information, then their effect would be indistinguishable from the effect of the CAM system. Such an effect, for example, can generally be obtained from the installation of numerically controlled machines.

The relative economics of direct computer control and numerically controlled machines for applications that do not require the handling of information (other than part programs) generally are prejudiced in favor of conventional numerically controlled machines. If one wants only an improvement of production rates and quality, this can often be achieved with conventional numerical controls. On the other hand, if improved information handling is wanted, the computer is most often the proper solution. If the manufacturing group were equipped with a CAM system which, for the sake of example, generated information that described the effect of its operations on the inventory of the storeroom feeding it, then this information would certainly be of some value to the working group that operates that storeroom. If the CAM system were equipped to handle bill of materials data, an order to manufacture a specific part could automatically generate an order to the storeroom for the retrieval of the materials required for the part. If the CAM system were to have access to the inventory files for that storeroom, it could represent this retrieval list in the form of a picking list that specified the locations within the storeroom for each required item. It might also modify the on-hand figures to account for the withdrawal of this material from the storeroom. Having this access to the on-hand figures, the CAM system might also monitor them for low-stock levels and generate reports to that effect which could be used by the purchasing organization. As a manufacturing group produces completed parts and passes them on to the next stage of the operation, it may generate storage lists that control a part of the operation of the buffer storerooms standing between the manufacturing group and subsequent manufacturing groups. If the CAM system has access to the inventory files for that storeroom, it can monitor the inventory levels in that storeroom. The measurement of these levels can be used to control the manufacturing group equipped with the CAM system. If stock rises above certain levels in the buffer storeroom, it may mean that the output of the manufacturing group is too great and should be cut back until the stock levels fall back to more reasonable values. It may also be the case that abnormally high stock levels in the buffer storeroom signal some difficulty in the next manufacturing operation with regard to its output rates or the manufacturing schedule under which it operates.

Thus we see that a CAM system operating in one group may have a significant influence on other groups if it has access to certain tactical information and utilizes this information in the proper way. The overall effect of the CAM system is a tendency to extend itself into other groups in the same vertical stratum and to organize and coordinate the activities of those groups in ways that tend to synchronize them with the group equipped with the CAM system. This effect represents the natural growth of the CAM system into the present manufacturing organization. It is, in fact, sometimes possible to initate the use of computer-aided manufacturing in a corporation by taking advantage of this natural extension of the system. If the system can be installed first in a single working group, it exhibits a strong tendency to grow into handling information that is of value to other working groups within the same vertical stratum. These working groups gradually find themselves utilizing this information and establishing their own links with the computer-aided manufacturing system. These links manifest themselves in several forms. They may, for example, involve extensions of the functions assigned to the original computer system. They may also involve the installation of entirely separate computers in the other working groups. These computers would then be equipped with data links between them, so that the exchange of information could be handled in machine-readable form and thereby carried out more efficiently.

Beyond the effect of handling tactical data and the resulting synchronization of working groups, the CAM system may also be expected to influence the management strata above the plane of the working groups. When we get to this stage, we are considering the problem of integrating the CAM systems into the corporation as a whole.

INTEGRATION INTO THE PRESENT CORPORATE STRUCTURE

The principal function of the members of management strata is to make decisions about the course of action to be taken by the working groups. These decisions are primarily based on information obtained from the working groups—information that measures their performance with respect to the goals of the corporation. The performance of the corporation as a whole is measured by statements of income prepared by the firm's accountants. Our next consideration is the effect of a computer-aided manufacturing system on the functions of personnel occupying these management strata. In what way can the installation of one or more CAM systems help these people perform their duties? We need to look at some of the ways that these people handle information in uncomputerized situations, and in situations in which whatever computers they may already employ serve as conventional business data processing systems.

In the uncomputerized situation, the manager must rely on the transmittal of information by word of mouth and by documents prepared by hand. At the first level of management strata—that occupied by the so-called "middle managers"—this information is cast as work-in-progress reports, low-stock reports, work orders, and so on. This information represents a first level summary of the tactical information being handled daily by the working groups. One of the first things that the middle manager does is to formulate this information into still more refined summaries for succeedingly higher management strata. One of his responsibilities, for example, may be the collection and correlation of data that represent the dollar volume presently invested in the materials kept in the storeroom over which he may have charge. If he is responsible for a section of the manufacturing facility, he may formulate these data into production reports stating the number of units that his group turned out over the duration of the last reporting period. Another use to which these data may be put is in daily operational decisions that fall outside the range of higher management responsibility. Such a decision might be to place a temporary hold on work for one customer so that work can be speeded up for another customer. This is a typical decision for managers of production facilities that are oriented to a job method of operation. The manager who makes these decisions must factor his knowledge of the customers involved in with the present state of his production operation, its capabilities, its projected work load, and other considerations.

In the conventionally computerized corporation, where the computer is used primarily for conventional business data processing, the middle manager is not assisted to any great extent by the existence of a remote computer installation performing only these processing functions. In fact, this computer may create extra work for the middle manager. Such an installation is generally operated under the aegis of the controller or treasurer of the corporation and its time is primarily spent in the preparation of payroll, issuance of checks, general ledger, statement of income processing, recording and summarizing transactions of sales and purchasing departments, and so forth. In each case, the base of data from which the computer works is representative in some way of the entire corporation, and the transactions that represent its daily inputs are prepared using conventional keypunch machines (or the newer key-to-tape and key-to-disk systems) by squads of keypunch operators who are part of the data processing department. The forms from which these data are punched are filled in by hand by the members of the working groups in which the transactions originated.

If the middle manager is responsible for one of these groups, a part of his duties may include supervision of the preparation of these data. In some instances, of course, the computer handles much of the clerical work that he might otherwise be responsible for—and in this case the business data processing installation can save the middle manager some work.

However, the tendency that frequently results from the automation of report generation at a high management stratum is an increased demand for more reports. This phenomenon is not unrelated to Parkinson's law in that the installation must be supported by a moderate number of persons who, if the same reports as were prepared previously by hand were all that were to be prepared by the computer, would frequently find themselves wanting for useful employment. Following the principle that work expands to fill the time available for its completion is the corollary that a low long-term average work load tends to create a demand for extra work to raise this average. Granted that there are exceptions to this rule (e.g., certain railroading operations), the data processing departments of corporations have not been found historically to be among these exceptions. Rather the case is that they or the departments to which they report discover (or invent) a need for additional information beyond that on which the corporation previously thrived. And so, the middle manager finds himself being asked to fill out an increasing number of forms that represent in various new ways the transactions of his working groups. The theory is that these new representations can result in more accurate and meaningful reports on the condition of the corporation—and this may very well be true, but the value of these reports is often lost on the harassed middle manager.

How might a computer-aided manufacturing system help the middle manager in such a situation? The answer to this lies in handling the information to which the CAM system has natural access, namely, the tactical data that describe daily operations. These data are, in fact, the transactional data from which the manager formulates reports and fills out forms. To assist this manager, the design of the CAM system must include converting this transactional data into forms that can be directly employed by the business data processing installation. This formatted data can then be expressed in some form of machine-readable media and transmitted directly to the business data processing computer, without the middle manager having to be directly concerned with its preparation. This form can be punched cards, magnetic tape, or direct communication between the computers over an in-plant telephone system.

The untilization of the CAM system can also benefit upper management strata. For example, as more data are cast automatically into machine-readable form, the load of work on keypunching facilities becomes lighter. The managers in these upper strata must be aware of the expansion of work principles that we have mentioned so that the proper action can be taken to reduce the number of personnel rather than invent still more data to be collected from working groups that have not yet been equipped with computer systems. Furthermore, the automation of this data gathering operation should result in faster data gathering and report preparation. Thus these managers benefit in that they receive their reports more promptly and the time lag in the control cycles over which they have responsibility is thereby reduced.

The question of where the computer can best fit into the corporation should generally be considered in light of the fact that many functions that the computer might perform are presently handled by people. Reasons for the computerization of information handling functions are generally based on the speed with which the computer can perform these functions and the occasionally reduced costs of preparing these data. There is generally a great deal of similarity between the techniques that the computer uses to handle these data and the techniques that people would use if they were handling the data. As we look to larger corporations, we often find, historically, that entire departments were devoted to data processing functions which are now handled by a single computer program. These departments still exist, for the most part, but their functions have changed from doing the data processing themselves to preparing the data for the computer. There are many analogues between the functions performed by the computer and the functions performed by individuals within an older form of corporation.

In very small companies, we may find that many of the functions are performed by a single individual. In a company with a modest number of employees, one clerk might reasonably be expected to handle all invoicing, accounts payable, and payroll. The installation of the business data processing computer creates a similar situation in the larger corporation. In the case of the corporate business data processing computer, we find that many separate departments and strata in the corporation are, in fact, time-sharing a kind of "super clerk." When the computer is processing payroll, which it may do once per week, it is serving one department. When it is processing accounts receivable, it is serving another department. Certain relatively recent additions to the software packages provided for business data processing computers include programs that analyze inventory and transactional reports which outline the condition of a manufacturing operation with regard to work in progress. These capabilities have had the commendable effect, in the instances in which they have been successfully applied, of improving the condition of the middle manager to some degree. But they occasionally fail to live up to the expectations that people have for them and they have generally been enormously costly in terms of development expense. In any case, the most important point is that all of the departments that have had their data processing automated to any extent find themselves sharing their computing facility with other departments. (The distribution of the corporate computer system over some major components of the corporation is illustrated in Figure 6.3.) Except in the very

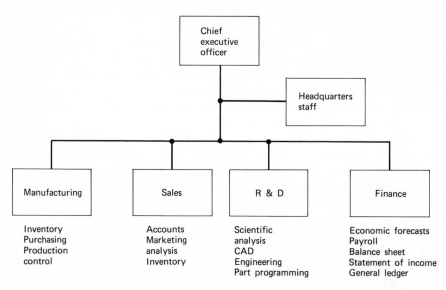

Figure 6.3. Use of the corporate computer.

largest corporations, the cost of large-scale business data processing installation generally prohibits a situation in which every department is furnished with its own computer. The computer-aided manufacturing system based on a number of relatively inexpensive minicomputers may very well be the solution to this problem.

We have identified this time-shared use of the business data processing installation as a problem. To what extent is this a problem? The principal manifestations in many cases are the delays that are incurred when one department must wait to receive the results of its processing while the computer is doing work for another department. Another is that the individual departments generally do not employ their own data processing personnel—system analysts, programmers, and so on. They must therefore rely on the services of these people through the auspices of the corporate data processing department. This reliance removes from the middle manager some of the control that he should have over the way that his data are handled. To lose this control without also displacing some responsibility for the results is not a desirable situation. Yet he generally finds that his is the ultimate responsibility for the way in which the data are handled; and frequent attempts to blame problems on the data processing departments are not always received in the most beneficent light by the managers to whom the middle manager reports.

Until the advent of the minicomputer, there was no reasonably cost-effective means of resolving this problem of reliance on the central data processing department. The situation now is that many companies are engaged in providing minicomputer systems specially programmed for particular business data processing that is directly related to the local problems of middle managers controlling working groups involved in manufacturing and related operations. These specially programmed minicomputers are available with software packages for the management of work in progress, inventory, purchasing, and even the generation of numerical control programs. The direct connection of these computers to computers handling direct machine control functions is a natural extension of the more basic forms of computer-aided manufacturing systems. It is expected that, through this approach to the problems of the middle manager, the CAM system will tend to extend itself up through the management strata of the corporation and eventually establish direct links with the original business data processing facility, which will continue to perform its traditional function of handling the large-scale data processing that serves needs that are common to all departments of the corporation.

ENHANCEMENT OF PRODUCTIVITY

Maximization of the return on investment should be the principal objective of the manager. We might therefore reasonably conclude that the primary reason that he might be interested in the installation of a computer-aided manufacturing system would be to improve his return on investment. The enhancement of productivity in the manufacturing organization is the first step for many corporations. In general, the first stage of computer-aided manufacturing is a small stand-alone support system for production control or inventory management or a modest system for the direct control of manufacturing tools. This first stage, as well as subsequent extensions of the system, must be justified by economic analyses of the present and future characteristics of the manufacturing operation, whatever its configuration might be. The system must be paid for from some present or anticipated pool of funds and it must be demonstrated, before these funds are firmly committed, that their expenditure will reap a satisfactory return. In fact, return from the investment of these funds should, in the expectations of the economic analysis, be greater than any alternate means of investing these funds within the charter of the corporation.

Theoretically, the correct way to determine accurately the proper method of investing from this pool of funds would be a complete cost-effectiveness analysis in which the goals are stated strictly as maximum return on investment. The actual situation is generally that the manager of manufacturing requests that his facility be provided with a CAM system. The practical problem of cost justification is then one of demonstrating that these funds produce *some* form of return greater than their present value. This return is usually expected to manifest itself in the form of improved *productivity*. To demonstrate firmly that this is indeed the case requires actual experience with the system, in place and working—a firmly established history of success. To get this positive demonstration requires actual investment of funds. Once the funds are invested, they are usually irretrievably committed.

If CAM systems were relatively inexpensive and generally available off-the-shelf, the company might persuade the manufacturer of the CAM system to allow them to try the system for a period of time at no charge. Unfortunately, except for a few systems that operate essentially as replacements for conventional numerical controls and require no customization, the manufacturer of the CAM system is both unwilling and financially unable to invest in this installation without some guarantee of return. The only remaining means open to the prospective purchaser of a CAM system is a theoretical study which may indicate the worthiness of investment in a CAM system. For these small-scale initial installations, the expense of a complete cost-effectiveness analysis might very well represent an appreciable percentage of the total cost of the system. The end result is that the analysis that is actually done is the simpler one of examining whether the system will produce any improvement at all in the productivity of the manufacturing operation.

And so, virtually all practical analyses of potential investment in CAM systems center around the question of improvement in present productivity. Figure 6.4 illustrates the concept of productivity that is employed in such an analysis. The productivity of the system is defined as some function which is generally monotonically increasing as the ratio of output to input increases. Complications of the definition arise from other factors, such as the condition of being overproductive, for example. This condition might arise when the manufacturing operation is producing more goods than the sales department can sell. We may also have this condition if the manufacturing operation is producing more goods than the distribution system can handle. In these situations, while the productivity function for the

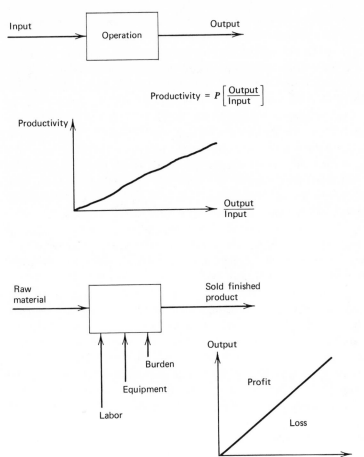

Figure 6.4. Concept of productivity.

manufacturing operation might continue to increase monotonically according to the increase in the ratio of its output to its input, we might also find that the inventory system is being inundated with manufactured outputs, that inventory carrying costs are going up drastically, and so on, with the total result that income for the corporation as a whole is declining, either as a percentage of gross or in actual dollars. The latter situation is very bad, of course. The former situation, although not so bad, may scare away potential investors and thereby deprive the corporation of capital that it may need to buy more CAM systems to aid the inundated inventory facility. In the simplest models, however, the improvement of the output/input ratio generally improves the productivity of the organization in question.

Cases of overproductivity are generally pathological and do not occur so frequently that they merit a great deal of concern.

The input and output of the facility are generally measured in dollar value or some reasonably convertible equivalent, such as number of parts produced over a certain amount of time. A time factor is also present in the productivity function. Whenever we consider the time differential of productivity, we generally find it to be nonzero. In some industries the cost of materials and the demands made on the corporation have seasonal variations. These introduce fluctuations in quantities that influence the output/input ratio and, therefore, the productivity of the operation. Although the electronics industry is not generally seasonal, a problem with time lag still exists. In the electronics industry, the principal contributor to the differential of the productivity is the installation of new equipment. The cost of a new system introduces a quantum jump in the input to the operation. Since no physical system exhibits zero time lag characteristics, some time must necessarily pass before the fruits of any operational improvements introduced by the presence of the CAM system are felt. If the immediate investment in the new system is too great, the corporation may go broke (in the extreme case) before improved productivity begins to replace the funds that were withdrawn from the pool of working capital. Rising labor costs are another contributor, but they can generally be statistically predicted to a reasonable degree of working accuracy using the figures that are collected periodically by governmental agencies. They must be considered, but the fact that they are generally rising is sometimes an argument for the installation of a CAM system, and ignoring them is a conservative error.

In performing an analysis of the effect of a new method (in particular, the installation of a CAM system), one must be careful to include all of the cost factors that contribute to the input of the operation. Among these cost factors, the contribution of the manufacturing operation to the burden (overhead) of the corporation must be included. This is particularly true for the manufacturing operation that is oriented toward job shop work, such as special systems, custom wiring, manufacture of customized printed circuit boards, and the design of LSI systems. It is also the case that an accurate measure of the value of the output is required. The remainder of the corporation must be capable of handling the output of the improved operation without degrading their own operation. The quality of the output is also a consideration. Producing finished goods at a greater rate may be counterproductive if the quality of the goods is lessened for any reason. This is a good example of a time lag effect. If quality falls, customer dissatisfaction sets in and the market for the goods is reduced. If this situation is allowed to continue, overall corporate productivity eventually falls off.

Now that we know what to watch for in evaluating the productivity of the operation and the corporation as a whole, we should consider how the computer-aided manufacturing system might be expected to effect changes in the output/input ratio.

At the lowest level of CAM system installation, where we are considering computerized equivalents of conventional numerical controls, the advantages are very much the same advantages that accrue to the use of conventional numerical controls. The use of part programs rather than manual specifications for a part improves the accuracy with which parts are made, and the increased speed of the controlled machines improves the turnout rate, generally by factors of three to five, or more. This is especially true of numerically controlled wiring machines, printed circuit board drills, and automatic component insertion machines. This speed and accuracy improves the degree to which personnel are utilized and reduces scrap and inspection costs. The use of computer methods, in generating the data from which the machines work, reduces the turnaround time for individual jobs and reduces the cost of machine setup, especially for wiring. As a result of this reduced turnaround time and the higher utilization of the working force, the manufacturing operation oriented to job shop work is generally able to handle the same volume of work at a lower expenditure of personnel and management time and thereby achieve a certain improvement in its productivity. To achieve this lower expenditure of personnel and management time means, of course, that fewer manufacturing and management personnel are required. Generally, the management personnel for whom requirements are reduced are clerical personnel who operate in support to managers in stafflike functions.

The principal difference between a conventional numerical control system and a CAM system, at the level of the manufacturing floor, is the information that the CAM system can naturally provide—the tactical information that we discussed earlier. Where this information was being handled with manual methods, the installation of a CAM system can make the handling of this information more efficient, either through the elimination of manual data collection or the use of direct computer-to-computer links for the transmission of this information to the corporate business data processing computer for final analysis and summarization. This causes a further reduction in the need for clerical staff personnel.

The information handled by this level of clerical personnel generally serves a control and stimulus function. Among their jobs, for example, are periodic examination of raw materials inventory and the generation of purchase orders when low inventory is detected. They also handle and record the effect of requests for material from the manufacturing operation, and handle the storage of semifinished and finished goods in buffer storerooms and distribution centers. The proper handling of this information requires examination procedures for detecting improper situations. For example, control over the generation of purchase orders is needed, so that two clerks do not generate orders independently of one another for the same item. It is sometimes difficult to obtain proper control over this, even by the expedient of assigning a single person to the job, since this person may have trouble recalling what he may have done a few days earlier and is therefore not immune to accidental duplication of effort. The CAM system that participates in the handling of this information can impose controls that help to prevent duplication of effort. It can also impose controls that force a coordinated discipline on the operation as a whole.

The principal improvement brought by the CAM system at the manufacturing level is in the handling of information and the degree of improved control that the system can enforce.

ENHANCEMENT OF EFFECTIVE CONTROL

Beyond the function of direct control of manufacturing machinery, the computer-aided manufacturing system serves principally as a support service and tool for the middle manager. It is most usually the middle manager who instigates the procurement of a CAM system. The CAM system operates under his control and extends itself into other areas of the corporation through the cooperative efforts of the middle managers who are most directly involved in the use of information provided by the system. To see how the CAM system can effectively serve the middle manager, we need to examine the issues of effective control over the manufacturing organization and the ancillary working groups that support it and distribute its output to the customers of the corporation.

In most manual systems and in systems that are conventionally computerized (in the sense of normal business data processing and production/inventory control systems supported by large-scale data processing systems), the information handling function is strongly centralized and removed from the direct control of the middle levels of management. Figure 6.5 illustrates some of the information flow for a rather simplified model of a manufacturing operation. The four principal functional elements of the organization are the following:

- A buffer store for incoming raw materials
- A manufacturing process that transforms these raw materials in some way
- A buffer store that holds the transformed (finished) goods
- A distribution system

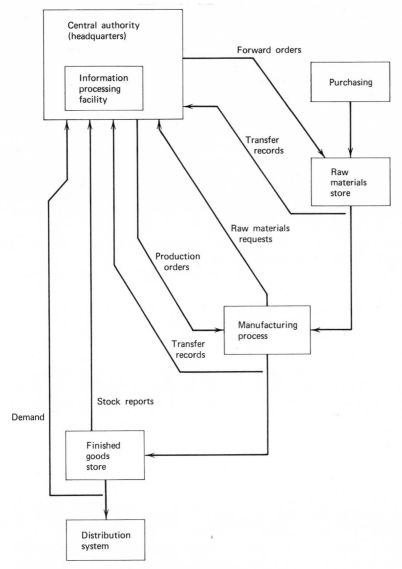

Figure 6.5. Centralized information handling.

As we have noted (e.g., Figure 6.2), the internal structure of these elements can be rather intricate. The central authority in control of this system may take on several forms. In the most strongly centralized form, all decisions and routine tactical forwarding of service requests from one working group to another are handled at a headquarters level. In a corporation in which the manufacturing facilities are geographically localized, all information processing and making of decisions based on this information may be handled in a single physical location. If the geographic spread of the organization is such that the logistics do not permit making these decisions in a single physical location, then the central authority may be fragmented and decisions made at points in closer physical proximity to the operations. These decision making points are still staffed by representatives of the original consolidated central authority. In handling the mass of tactical data, where procedures may be established for the control of normal situations, control functions are often handled by personnel who function, in effect, as a staff to a higher level of management.

As the corporation grows in size and geographic spread, this fragmentation continues. As a member of this fragmented central authority, we find the middle manager responsible for the operation of some working group. However, the fragmentation of the central authority is not always accompanied by a corresponding fragmentation and spread of information handling facilities. Thus, as the corporation grows, the middle manager finds himself further removed from the sources of information that he needs to control his operation properly. The information channels, because of the concentration of the information processing facilities, carry demand measurements from the distribution system and stock reports from the finished goods storerooms to the data processing facility. There it may be discovered that manufacturing output must be either stepped up or down. This results in a sequence of production orders coming back down to the manufacturing facility which, in turn, create demands for more or less raw materials. These demands must be reported to the information handling facility where they are converted into withdrawal orders from the stock of raw materials and purchase orders for the replenishment of the withdrawn materials. Thus the chain of information transfer and processing is considerably lengthened and the middle manager is removed both from prompt access to the information and from control over the information and its forms of presentation. Occasionally, because the manager has no control over the information, and no say in the analysis of the information, he may find that a good deal of the information is of little value to him. He may also find that he cannot use the information to enforce operational disciplines on the working group for which he is responsible.

If the flow of information could be revised to conform more closely

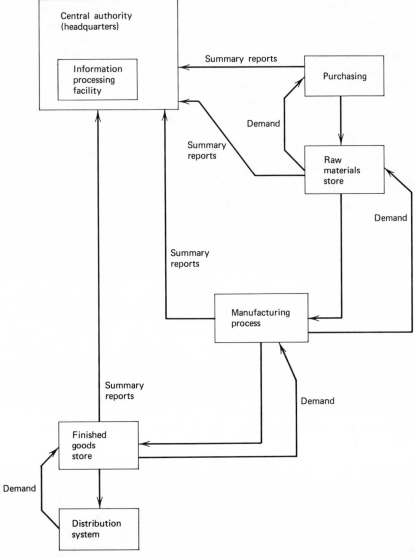

Figure 6.6. Decentralized information handling.

to that illustrated in Figure 6.6, we might expect the situation to be somewhat improved. In this figure, we see channels of primary information flow that conform more to the requirements of the middle manager. As he controls the operation of his working group, he must make demands on the element that feeds his group and pass on something to the group that he is feeding. The demand presented to him from the group that he feeds should act as a control on the material that he passes on to that group and on the demand that he places on the group that feeds his group. Thus we would have a situation in which local authorities were handling local data for the control of the groups for which they were responsible. The manufacturing operation manager could analyze his own raw material requirements quickly and generate withdrawal requests directly

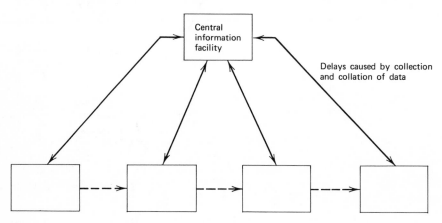

Figure 6.7.   Centralized information handling.

to the storeroom. The amount of finished goods absorbed by the distribution system can directly control the finished goods warehouse, which can pass its requirements directly to the manufacturing facility. Properly condensed summaries of these activities can then be passed directly to the central information handling facilities for analysis of the performance of the corporation as a whole, the proper function of the centralized information handling facility. The overall effect is to improve the middle manager's control, to reduce information traffic, and to improve coordination between the working groups.

To abstract these concepts somewhat, we turn to Figure 6.7. Here we have a sequence of individual subsystems in which material is passed from one to the other. For the proper operation of this collection of individual subsystems, it is necessary that the flow of material be controlled according to certain rules. In the case of the centralized information handling facility, the information on which the control decisions are based is collected and transmitted to a central facility. There it is collated, analyzed, and formulated into reports that indicate the nature of the decisions which must be made to effect the proper operation of each individual subsystem and the collection of subsystems. Long delays are caused by the preparation of transmittal media, the fact that the central facility is being shared in order to satisfy the processing requirements of all of the individual subsystems, and the transmission of the analysis back to the middle managers

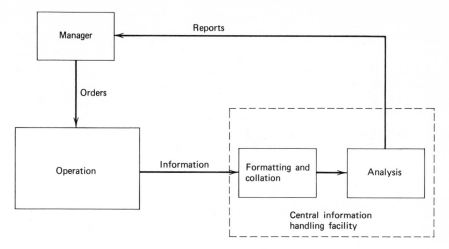

Figure 6.8. Centralized information handling.

so that they can take action on the results. The manager's effective control over his working group is attenuated because of the long time lag between an event, its detection, and eventual correction. Because the manager does not have direct control over the information handling, he cannot tailor the information to his particular needs and the needs of his working group. In Figure 6.8, we see illustrated some of the detail of one loop in the centralized information handling scheme.

In Figure 6.9, the information handling function has been decentralized and each manager exercises direct control over the information most pertinent to his own working group. The manager is directly concerned with only two channels of material flow—the one coming into his operation and the one going out. He is concerned with the generation of a single demand function—the one that serves to control the input to his opera-

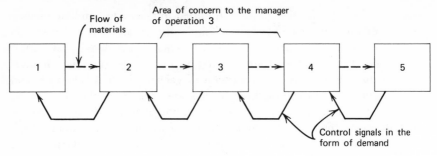

Figure 6.9. Decentralized information handling.

tion—and he must respond to only one demand function—the one that serves to control the output from his operation. Having his own information handling operation means that the manager can exercise tighter temporal control of the information that pertains to his working group and that he can also exercise control over the handling of this information in terms of priorities and techniques (Figure 6.10).

It is frequently the case that a quick survey of the present state of the working group is of great value in planning and controlling its daily opera-

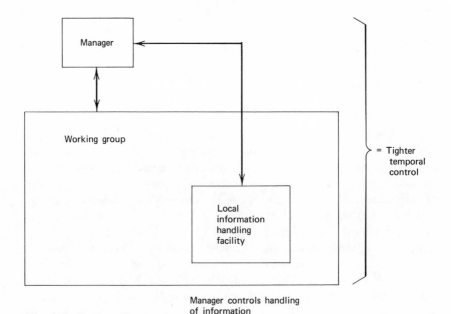

Figure 6.10. Decentralized information handling.

tion. In the job oriented manufacturing facility, for example, the manager would like to know, before formally starting work on a particular job, whether all of the required raw materials and semifinished parts are on hand. He may also wish to check work in progress to see if the personnel and machines that he might assign to a new job might not be better employed assisting in the completion of some other job that is behind schedule. The manager of a warehouse or buffer storeroom would like to be able to reserve materials for a withdrawal request, pending the actual picking of the items, and to manipulate the way these items are reserved so that he can coordinate the restocking and withdrawal process with the manufacturing process.

If an examination of data that told the manager the situation of his working group could be made promptly, then he could make procedural alterations more effectively. If there is a large time lag between the actual measurement of the state of the system and the report of the measurement to the manager, then the data are no longer valid except as a measure of overall performance. If the data are timely, then the manager knows the state of the system at very nearly the present moment. He can see what resources are actually available at that time for reallocation, whether stock is truly on-hand to meet a certain withdrawal request, and so on.

The manager can also use these data to impose normal procedural disciplines on the operation of the working group—disciplines that would not be possible without timely information about the true state of the system. For example, in a storeroom feeding job shop assembly operations, one should not withdraw parts to go into a particular subassembly until all of the parts for it are available. If some parts are not available, the complete assembly cannot be made. The partially finished assembly would have to be put into temporary storage to await the missing parts and the smooth flow of work would be disrupted. Ideally, work should not begin on an item until everything is ready and the full cycle of the manufacture of the item can be accurately predicted. There are also occasional situations in which the manager would like to exercise special control over the sequence in which things are to be done, so that he can meet certain special schedules that may be imposed on him. In the majority of the situations, the manufacturing operation performs work as it is received—this is usually the desirable way to operate the system, since in this way it has a tendency to run itself when the situation is normal. But the manager must be able to modify the operation of the facility when unusual situations arise. With local information handling capability, the CAM system can perform some of the scheduling function automatically. For example, working from a list of work ordered in the way in which work orders were received, the CAM system can check availability of raw materials and automatically postpone work for which materials are not available. It can, at the same time, monitor the status of the materials in the storeroom and automatically reinsert the work in the sequence when the parts are finally available. In this way, the CAM system can relieve the middle manager of some of the daily work of sorting out the order in which work can be done. The CAM system then provides the manager with reports on how it has laid out the work and the reasons behind its automatic response to any unusual circumstances. If the manager wishes, he may override the decisions made by the CAM system. If these decisions were to be handled by a centralized information handling operation, the time delay between gathering of the data, analysis by the information handling system, and transmittal of reports and manufacturing orders would lead to serious inefficiencies in the operation of the system. By the time orders came back to the manufacturing floor, the situation might no longer bear any resemblance to the situation that generated the data which lead to the new orders.

Until very recently, the situation was such that localized handling of information to this extent was not economically feasible. The principal reason for this was the large cost of a proper computer installation, the cost of personnel to maintain and operate the installation, and the cost of preparing the software to do the data processing. The advent of the minicomputer has brought several changes to this situation, particularly to the cost of the computer hardware. For many small and medium manufacturing operations, the required data processing capabilities can generally be provided by a minicomputer with a modest collection of peripherals, such as a Teletype, a disk, and perhaps a card reader. For many of the data processing problems that are faced by small operations, the development of the software can be handled by a relatively small staff of programmers who need not be retained as a permanent supporting staff.

Proper grouping of responsibility and the effectiveness of the manager can also be improved through the utilization of the information handling capabilities of the computer-aided manufacturing system. The objective behind proper grouping of responsibility is to ensure that the manager who is responsible for the control of an operation is also responsible for the gathering and processing of information about that operation. Placing responsibility for information handling on organizations outside the direct control of the manager makes it possible for him to attempt to shift responsibility for substandard performance to the information handling operation. In some cases, he may be completely justified in this. As we have noted, the centralized information handling facility must be shared among all of the operations, and the data collection process is often somewhat cumbersome when the handling facility is so far removed. As a result, the data that the manager does receive are often of little value to him because they are out of date. It is neither his fault nor necessarily the fault of the information handling facility, for the facility may have processed the data just as fast as it could. The fault lies with the way the handling of information has been set up. In some cases, however, the information handling facility's long delay time may be a convenient scapegoat for the manager who does not actually have adequate control over his operation. The problem for the next higher management stratum is the arbitration of the dispute and the determination of where the fault truly lies. This is often a difficult task.

Perhaps we should take a moment to emphasize that the reasons for giving the manager control over the handling of data for his own organization do not include giving him the ability to improve falsely performance figures for his working group. The measurement of performance figures, if considered from the viewpoint of measuring profit, is more than adequately covered by the accounts. Furthermore, production rates for groups are measured at the working group interfaces—what is output from one working group is input to the next working group in the sequence, and a cross check therefore exists at these interfaces. The difference between the demands generated by one working group and the output of the working group that feeds it is a measure of the performance of the feeding work group. The difference between the demands that the working group makes on its feeder working group and the demands made on it by the group that it feeds in turn are a measure of the efficiency with which the group converts demands into actual production of goods. Thus even with local control of information handling operations more than enough data are provided to the higher levels of the corporation to enable them to measure accurately the performance of the individual working groups and the degree to which they cooperate with one another.

The effectiveness of the individual manager is very much a function of the amount of responsibility that he has and the amount of information that he must process in order to execute properly the functions required to carry out these responsibilities. Figure 6.11 illustrates the behavior of effectiveness as a function of information and responsibility. The implication of this figure is that there is an optimum value of both responsibility and information handled at which the effectiveness of the manager is maximized. If the manager has too little responsibility, he cannot adequately control his operation. If he has too much responsibility, there may simply be too much for one man to handle. If he has no information about the operation, he has nothing on which to base his decisions and they are therefore probably wrong. If he receives too much information, it may be too much to digest. He may become inundated with information and be unable to recognize conditions and trends that merit his attention and action. The

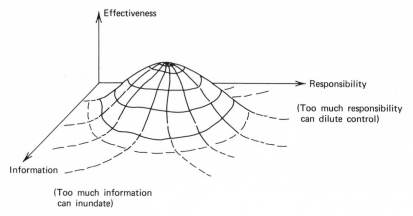

Figure 6.11. Effectiveness as a function of information and responsibility.

detailed shape of this effectiveness function is determined by the ability of the manager as well as the characteristics of the working group over which he has charge. It is also influenced by the nature of the information handling system that the manager uses in making his operational decisions.

The computer-aided manufacturing system can be used to improve the effectiveness of the manager through its influence on the information handling operations. By using the CAM system to filter information, the possibility of inundation is reduced. If the manager's present situation happens to place him on the side of the function nearest the origin, placing control over the information handling facility in the hands of the manager can give him the degree of responsibility needed to enhance his effectiveness, giving him tighter control over his organization through the effects that we have been discussing, such as better temporal control over his working group.

There are many different degrees to which the computer system can participate in the manufacturing operation and its management. The uses to which the computer is put, and the degree to which it exercises control over the normal sequence of daily operations, are functions of the amount that the corporation is willing to invest in the installation and the way the corporation views the projected return on this investment. It has much to do with the lifestyle of the corporation and the way the corporation views itself internally and in relation to its competitors.

The lowest levels of computer participation in the manufacturing process and the management of the process are at the local level, where the computer either acts as a direct controller of machinery on the manufacturing floor or as a clerk to support the middle level managers who are directly responsible for the production operation and its supporting services. These are the easiest levels to install and they generally produce the most immediate rewards in increased productivity and operational efficiency. Furthermore, computer-aided design installations operating in the engineering groups of the corporation can feed data directly to these machine control operations. At the next higher stages, we find a linking of these separate computer systems into a more coordinated system. For some corporations, these are represented by some of the large direct numerical control systems that have integrated the functions of part design, machine control, and production control into a single system. For the most part, such systems are used in the metalworking industries and are now being sold by some of the more advanced machine tool builders. The concept of the integrated electronics manufacturing systems that would be analogous to these have not yet appeared on the market as off-the-shelf systems, possibly because few corporations are in the position of manufacturing machine tools that cover the whole gamut of electronics manufacturing machines. At the highest level, in the largest corporations, we may find a large-scale data processing installation on which major production and inventory control programs are operated. These same computer installations may be employed for simulation studies and long-range economic and market forecasts and therefore might be considered as supporting the corporate philosophers (long-range planners). The most pressing questions of computer aids to manufacturing are relatively removed from such computer applications.

One of the most important principles that must be kept in mind in setting up one of these systems is that of maintaining flexibility. When the manager of a manufacturing installation is provided with his first CAM system, it is going to be unfamiliar to him—even if he has studied its capabilities very carefully and assisted in the establishment of its design specifications. There is little question that some changes will be made to the system during its lifetime—either as a result of new understanding of old problems or from the development of new methods of operation and data handling. This should be considered in the design of the system and the establishment of the procedures under which it will operate. It may also be required to change its method of operation so as to incorporate communication channels to other computer systems in the organization. The most general way to maintain flexibility to adapt to these changes and new methods is to maintain a reasonable degree of modularity in the design of the software for the system. The software should be organized around the methods that the system will be implementing initially, with a strong association between critical functions and software modules. With this approach, it is much easier to locate and modify that portion of the software system that must be changed in order to implement a new policy or generate a new kind of report.

ACCEPTANCE OF THE SYSTEM

The acceptance of the computer-aided manufacturing system by the personnel who must use it and work with it, either directly or indirectly, is an extremely important factor in the ultimate success of the system and the degree to which it accomplishes its design objectives. This acceptance does not come to pass on the first day of system operation. Indeed, the system may never be fully accepted unless its design and method of operation are set up to solicit the cooperation of its users and other surrounding personnel. Thus the human factors of the system are very important. We have discussed some points in this area (Chapter 5) with regard to systems used for direct machine control. We should mention here that the systems that are used primarily for local information handling functions are somewhat constrained in terms of the traditional human factors considerations, since their designs are essentially dictated by the requirements of the data processing problems. That is, the system is generally nothing more than a small computer system and its interface with people is somewhat constrained by the information that they must exchange with it. Of course, the usual caveats apply to the design of printouts and operator communication with the system—they should be designed for easy and unambiguous use. However, the way the system is presented to these people is very important. It must be presented to them as an assistant to help them with their duties, and the system should be explained to them in the context of their duties as the personnel understand them. The fact that one of the economic motivations for the installation of such a system is a reduction in labor requirements will not make this presentation easier—many employees will, with good reason, view the system as a threat to their jobs.

Beyond the local acceptance of the system by the personnel who will be working with it, we have the acceptance of the system as a valuable element in the manufacturing organization with respect to improvements in productivity. Here, the system must be viewed from the perspective of an objective measurement of its effect on the profitability of the corporation. This effect will be borne out over a period of time through a very complex existing system of people, procedures, and other equipment, and it will take some time for the full impact of the system to be measured accurately. Thus an essential element of the integration of the system into the corporation is the ability to measure its performance against the goals of the corporation—goals that may change over the period of the design and installation of the system. As we stated above, the CAM system must remain flexible so that it can react to the changes in these goals.

To ensure that the system meets the goals of the corporation and is properly integrated into the corporation, care must be taken in getting started—the subject of our next chapter.

Chapter Seven

Getting Started in Computer-Aided Manufacturing

In this chapter, we discuss some of the managerial aspects of planning, selecting, and procuring computer-aided manufacturing systems. Because it is intimately involved with complex technologies and deals with a rather extraordinary segment of the business and industrial community, the subject is not exactly a straightforward exercise in the application of the traditional principles of good management.

We examine some guidelines for planning CAM systems, decisions regarding the kind of system to install, planning for the expansion of the system to meet the growing needs of the corporations, estimating the potential contribution of the system to the manufacturing operation, and justifying the system on the basis of its potential contribution. We then discuss procurement of the system, the selection of vendors, and the controls that should be included in the procurement cycle to provide the best chances for a successful system.

PLANNING

The reader is surely aware from reading earlier portions of this text that a computer-aided manufacturing system is a highly complex example of current technologies in digital systems, computers, numerical controls, mechanical design, and production management and engineering. Computer-aided manufacturing systems influence every aspect of the manufacturing operation in a very direct way and can have important effects on the operation of material procurement systems and inventory management systems in the corporation. Because the system represents both present and potential influences on several segments of the corporation, its installation must be carefully planned. Failure to recognize these influences and to plan for the control and proper utilization of these influences can be quite costly. The costs can manifest themselves in terms of the correction of situations that are detrimental to efficient operation and in the loss of desirable services that a properly conceived system could have performed had the need for these services been recognized early enough in the planning and installation of the system.

To plan a system properly, one must be aware of the options that are available in the selection of the system, its installation, and its future expansion. Therefore a necessary prelude to the conception of a concrete plan is a thorough survey of the available options. This survey must provide acquisition of the necessary technical knowledge on theory of operation and methods of application, awareness of the kinds of problems that are most frequently encountered in the installation of computer-aided manufacturing systems, and familiarity with the products that are available from potential vendors of all or any part of the CAM system. The results of this survey should enable you to benefit from the experience of others. It should lead to an intimate knowledge of the application of computer techniques to your own manufacturing system and may give you some ideas that you might not have had otherwise.

There are five principal sources of information for this survey activity. The basic technical knowledge can generally be obtained from textbooks on the subjects of digital systems and numerical control techniques. There are many such texts available and they will have something to offer to the development of the necessary foundation of technical knowledge. The remaining principal sources of information are trade magazines, trade conferences, actual installations, and vendors of computer-aided manufacturing systems, components, and subsystems.

Trade magazines frequently carry articles that describe actual installations, techniques that have application to the computerization of manufacturing and production control operations, and surveys of currently available systems and components. As a part of the survey activity, the annual indexes and recent editions of these magazines should be reviewed for articles pertinent to the subject. These articles should be obtained, studied, and retained for future reference. The principal vendors of CAM systems, and the components and subsystems that make up CAM systems, generally advertise in these trade magazines. These advertisements are a source of the names of vendors with whom you may become involved in the procurement of your system and generally provide useful information about the products offered by these vendors. The reader service cards usually included in these

trade magazines are a convenient vehicle for obtaining more detailed literature describing these products and their costs and relative merits. They
should be used and a reference file containing this literature should be
built up. These magazines also make it a point to preview significant trade
conferences and generally list the papers to be presented at these conferences, often with brief abstracts of the contents of these papers.

If the potential buyer is armed with a technical knowledge of the problems and theory of operation of CAM systems and with a general knowledge
of the nature of the CAM-related products offered by vendors, the trade
conference can be beneficially used as a further source of information.
Trade conferences that bear a relation to the topic of computer aids to
manufacturing usually have two parts: demonstrations of products in an
exhibition area and sessions in which papers are presented and technical
subjects are discussed. These sessions present the results of actual experience
on the part of users of computer aids and often provide valuable insights
into the installation of computer-aided manufacturing systems and their
design and utilization.

One can generally evaluate the desirability of attending these conferences
by reviewing the list of exhibitors and the abstracts of technical papers
and sessions that are to be presented. These lists and abstracts are printed
in several trade magazines several months before the conference is held
and they should be studied carefully. If it is not possible to attend the
conference and the presentation of the technical papers and to meet the
authors, an attempt should be made to obtain a copy of the proceedings
of the conference so that the text of the papers can be read. It is also
advisable to make personal contact with the authors of papers that appear
to be most applicable to your own planned CAM system. This contact
may be by letter or telephone, if not in person. Usually, these authors
have had valuable experience and are proud to be asked to share this
experience with others who are planning computer-aided manufacturing
systems.

The exhibitions at these conferences give you an opportunity to see the
components of CAM systems and, on some occasions, to see complete systems operating on a small scale. The opportunity allows you to make an
initial evaluation of the general quality and usefulness of certain systems
and to collect more information about them. The competing systems are
in close proximity to one another, so that you can crosscheck initial impres-
sions of relative merit with repeated examination of the hardware in a
short span of time. These conferences generally run for two to three days.
This provides you with time to study the systems, mull over what you
have seen, and return to reexamine the systems and ask further questions.
Most corporations exhibiting at these conferences make it a point to have

knowledgeable technical personnel in attendance so that you can set up appointments to discuss technical details and obtain answers to questions about the detailed nature of the equipment and its application to your particular problem.

These exhibitions have much to offer, but they are by no means the complete answer to the information gathering problem and they do have certain drawbacks. It is quite difficult to carry on a comprehensive technical discussion in the middle of a crowded exhibition hall; so you should try to arrange separate meetings with the exhibitors' technical personnel, if possible. Usually, they are not inclined to meet with you unless your corporation is actively considering a procurement or has actually committed funds to the project. This is an understandable position—they are there to sell their products and they have no motivation merely to inform the casually interested engineer or manager, unless there is a real chance of selling some of these products. Therefore, unless you are actually in the position of being a probable buyer, do not be put off by a lack of interest on their part in spending a great deal of time with you. It is also the case that there is frequently a good deal of "hucksterism" at these exhibitions and one can contract a good case of mental or sales fatigue from this aspect of the show very quickly.

The principal advantage of the trade conference is that many people of importance, along with the best examples of their products, are gathered in one place to aid in the solution of your problems. This makes it possible to compress a great deal of work and information gathering into a very short time.

One very important item of information that you should collect as a part of your information gathering activity is a list of installations of equipment in actual manufacturing operations which you may be able to visit and study for yourself. Unless they are in your competitors' facilities, you generally find that the owners of these installations are happy to show off their systems and freely discuss any problems that they may have had and the degree to which the systems are working out as planned. The vendors of the systems can provide you with the names of the people to contact, the locations of the installations, and may offer to arrange a tour for you. The success of the vendor in arranging such tours is an indication of the satisfaction of his customers with the equipment, and you should certainly allow the vendor to arrange some of the tours for you so that you can make this measurement. However, the vendor is not inclined to bring you in contact with his less satisfied customers, and unless you make some of the contacts directly and without the involvement of the vendor, you may find yourself seeing only the good points and never discovering some of the problem situations. You may also find it advantageous to contact other potential users of the vendors' products, so that you can compare notes with them about the relative merits of competing systems.

Depending on the relations between your corporation and its competitors, you may be able to get some valuable advice from them. The electronics industry is unusual in that such information is more freely traded than in many other industries. While your competitor will not reveal his most intimate secrets, he may very well be willing to share his opinions on the application of computers to your common manufacturing problems and on the vendors of systems for this function.

Aside from trade conferences and exhibitions, the vendors are excellent sources of information. You should get to know their representatives in your area and stay in contact with them. The representatives of computer manufacturers are particularly useful in this regard. They come in contact with a wide range of computer applications and, if they are good in their trade, they know a great deal about the problems of computerizing manufacturing operations.They are also a good source of contacts with local consultants and firms with experience in the actual development of these systems.

The information gathered during your survey activities should enable you to plan the full scope of computerizing your manufacturing operation. By the time that you believe you are ready to terminate the formal information gathering phase of your preparation, you should have a good intuitive feeling for the way in which computerization applies to your manufacturing process and should be ready to lay out a detailed plan.

This plan, when it is complete, should include the appropriate provisions for the computerization of the operation to the fullest extent that you can foresee. Overall planning for a fully integrated CAM system is important to the ultimately successful operation of even the more modest systems. It is clear that this is true if your corporation intends to pursue seriously the automation of its manufacturing operations to the fullest extent that can be found economically desirable and feasible. However, what if you are interested only in the improvement of a single segment of your operation and cannot presently visualize any significant probability that further computerization would ever be required? Do you then need a plan? The response is in the affirmative for, if you can logically deduce from your knowledge of your business and its technologies and markets that the requirements for such computerization will never come to pass, then you already have a plan. This plan may very well forswear the efficient expansion of computer aids beyond those that you are presently considering. In any case, if there is any reasonable probability that you will wish to extend computerization beyond the first stage, then it is wise to plan the complete process so that all of the elements of the system fit together with the least cost and difficulty.

If, on the other hand, you are presently convinced that complete computerization is indicated and you intend to pursue the plan to this extent, then you are faced with the possibility of full automation. Generally, the instant incorporation of a complete, fully integrated, system is inadvisable. It is better to implement the system through a phased plan in which the complete system is installed over a period of calendar and operational schedules that allow you time to use and evaluate modules of the complete system in relatively independent sequential stages.

There are several reasons for this. First, negative impacts on your present manufacturing volume should be minimized. One should not leave himself vulnerable to the possibility of extended lapses in the output production stream. Such lapses can come about from unforeseen problems in installation, schedule slips on the part of subcontractors, and many other events. Localizing the installation of computer aids and implementing them in modular steps help you and your vendors to control those aspects of development that can lead to these problems and minimize their impact if, for some reason, problems do arise. Second, every corporation has its unique problems. Some of these problems may be unforeseen and may not become apparent until the system has been in operation for a period of time. You should, therefore, also allow yourself some time to evaluate the performance of each part of the system as it is installed in each phase of your plan. The results of these evaluations can be employed profitably to make adjustments in subsequent phases of your implementation plan or to make modifications to portions already installed so that they fit more efficiently with subsequent phases.

The first phase of your installation plan should be selected on the basis of your evaluation of where the most immediate benefits may be obtained with the greatest probability of success. Except in those rare instances where the complete corporate management is convinced of the need for computer aids, these immediate benefits may prove to be valuable ammunition in obtaining the authorization to proceed with the implementation plan. Even with a management that is fully convinced at the start, problems with initial installations may temper their convictions at a later time. It is very important to have the cooperation of the management, especially with regard to the authorization of funding and the delegation of the proper responsibilities and controls to the managers directly associated with the computerized systems. Without this cooperation, the usual problems encountered in carrying out the implementation plan can be exacerbated or become insurmountable.

The starting point for your installation must also be selected in the light of the amount of present resources that can be committed to the project. The available funds will obviously constrain the first choice, but you should be sure that you do not overcommit these funds by devising a plan that is too ambitious in its initial phases. The principles of handling these funds are similar to those that apply when you are starting your own business: you should be sure to obtain more than you actually believe you need. Budget overruns have occurred with alarming frequency in the history of automation projects, often for causes beyond the direct control of the responsible managers. It is better to have a surplus in the budget than to have to come back for more money later in the project. This will cast a pall over the general impression of the project in the minds of upper management levels. Thus if you are unable to obtain the funding and surplus that you believe to be necessary, it would be wise to moderate your plans to create this surplus after the funds have been authorized. Your plan should provide for a review of the available budget after the initial approval of funds so that you can be sure that adequate surpluses are available in the event that difficulties do develop.

Another principle to follow in setting up the beginning phases of your implementation plan is to constrain yourself to the best established technologies and components with the most proved histories of success. This helps to alleviate some of the problems that may come to pass from the utilization of new devices and components. However, this consideration is a difficult and sensitive one. There will be times when the best component or subsystem for the job may be relatively new on the market and the use of an older model of the device may entail the inability to include some function that you want very much. Here there are difficult decisions to make and they must be made in the light of the present circumstances and requirements for the system. An important aspect of this problem is the general trustworthiness of the vendors. One will find situations in which a new company is competing with a more established company for the sale. At first glance, one might presume that the older company is the more trustworthy; especially if they have placed a lot of their equipment in similar installations. However, one may also find that the older company is not as well motivated as the younger one in terms of providing the best service. Here, you must also evaluate the importance of your business to your potential vendors and the degree to which they agree with you in regard to its importance to them. Price is also a consideration here. The newer and younger products may be more expensive than those with longer histories. The reasons for the price differences should be investigated, for one sometimes finds that the differences are reflected in provisions for better services, components of higher quality and reliability, and better features.

The general rules in this case call for keeping in mind all of the principles of dealing with suppliers. These include caution in evaluating their claims for their products, understanding their motivations, and evaluating their ability to give you complete satisfaction and their full intention to do so.

An extremely important aspect of planning is the selection of personnel for the implementation of your system. Men with the proper collection of talents for implementing and managing ambitious automated systems are not easily found. If you do not already have such individuals on your staff, you should begin the search for them as early as possible. It would be beneficial if individuals who participate in the planning, procurement, and installation of the system could become the final operational managers of the system. They will require familiarity with the technical principles of the operation of the equipment (so that they understand its capabilities), the management of electronic manufacturing operations, the nature of the market for the output of the manufacturing operation, and the best ways in which the system can be managed in order to improve the economic situation of the corporation.

In some corporations, there may be political and logistical problems in placing one individual in charge of all aspects of the CAM system. This is particularly true with large-scale, fully integrated, systems that influence virtually every department related, however remotely, to the manufacturing operation. In these cases, it is generally expedient to involve several managers in the initial stages of the implementation, with a single manager in overall charge of the implementation up to the point at which the system is so widespread that single-point managerial control is no longer effective or desirable. Fortunately, the principles of decentralization (which we discussed in the previous chapter) and the advantages of decentralized computer controls make such an approach feasible. With a properly phased implementation plan, this state should be reached through an orderly progression of events. If the plan is well conceived, it should come about in a natural fashion; the decentralization effects of the CAM and management information systems and the growing familiarity on the part of all managers involved with the system enable each manager to take over control of his own operation in a smooth and orderly fashion.

If you are planning a fully integrated system, your own unique needs may have the greatest single influence on the design of the system. Fully integrated systems are not available off-the-shelf and the areas in which they will require the greatest deal of custom tailoring are the software and procedural aspects of the system. These aspects of the design are the major work of systems analysts and computer programmers. One early decision that you must face is whether to develop the system yourself or to subcontract its development. If you decide to do the development, you will have to build up your own systems engineering and programming staff. Unless the future plans of your corporation include their type of work, they will necessarily be transient workers. The acquisition of such a staff is not a straightforward task.

If you do subcontract the development of portions of the system, you may find yourself in need of technical advice and assistance in monitoring and controlling your subcontractors. Depending on your location and the scope of the system that you are planning, it may be possible to obtain this assistance from a third outside firm with experience in the areas of system engineering and software design. The Defense Department occasionally adopts this approach in the development of large-scale systems. They contract with a third firm for monitoring of the work performed by the major contractors on the system. In the realm of manufacturing system automation, the analogue would be to retain a firm of consultants having experience in the area or to obtain assistance from a systems development company or software firm with appropriate-experience in the development of CAM systems. They can provide valuable advice and guidance and, if their responsibilities are properly allocated, they can effectively monitor the contractors that make the major portions of your system. Such an approach is advisable even if you have established a single prime contractor with complete responsibility for the complete development of your system.

You must be sure to establish the proper controls over your monitors, as well. For example, if they are essentially in the same business, either directly or indirectly, as one of your subcontractors (e.g., a software firm handling your computer programming), it should be made clear that there is no way for them to expand the scope of their contract to win work away from their competitor.

COSTS

Another important factor in the planning and implementation of a computer-aided manufacturing system is the element of cost. In the ultimate analysis, the most critical argument for the system is the economic benefit that it can bring to the corporation, especially in terms of improved profitability of operation. The decision to install a CAM system is usually based on indications that such benefits will actually accrue. Once the decision has been made, subsequent investigations and design decisions must be such as to assure that this is, indeed, the final situation. Having decided that CAM is a viable manufacturing technique for the corporation and having decided to investigate the kind of system that is required, the plans must be laid with these economic considerations in mind. There will be variations on the system design and the selection of components that will affect the total manufacturing costs. Some of these variations will be more economically attractive than others. The system planner must devise his implementation plan and phasing and select the subsystems and components that make up the system so as to take maximum advantage of economic factors and to maximize the return on the investment represented by the funds allocated for the system.

As with any kind of control system (physical, mechanical, or economic), accurate measurement or estimation of the system parameters is paramount in importance. The system cannot be controlled if it cannot be observed and a decision cannot be made regarding courses of action unless the quantitative results of each course of action can be compared. In the situation at hand, we have a present manufacturing system and several computer-aided manufacturing systems from which to choose. In making the initial decision to employ CAM, it was necessary to measure accurately the true costs of operating the present manufacturing sytem and to predict accurately the costs of operating the corresponding CAM system. In choosing among variations on a CAM system during the planning stage, it is also necessary to predict accurately the operating costs of the CAM system in each of its variations. The problems of economic measurement and prediction are severe and require an intimate knowledge of the economic behavior of many segments of the corporation beyond the actual manufacturing operation. Overhead considerations, the future availability of capital, the ability of the market to absorb the manufactured output, and so forth, are all considerations that must be made in one way or another in the choosing and planning of computer-aided manufacturing operations. For these reasons, it is important to utilize the services of corporate accounting personnel in setting up plans for the system.

As if the problems of economic measurement by conventional accounting rules were not sufficiently difficult, the CAM system planner must also consider a number of intangible factors, some of which can be considered to have genuine economic worth which is generally not measureable. Among these factors are the relations with employees before and after the installation of the system—how the employees will react to the system. Another factor is the goodwill of customers. The system may yield products of higher quality and thereby engender a higher level of customer goodwill. The assignment of a definite economic quantitative worth to these is virtually impossible, yet they must be considered when planning the system.

We cannot hope to offer comprehensive programmed methods for the economic analysis of CAM systems and the comparison of their economics with other kinds of systems. Each corporation has its own methods for determining the worth of the various factors involved, some correct and some not so correct. Many corporations do not, in fact, have an accurate quantitative measure of their present manufacturing costs. Corporations in this position generally find themselves in less than desirable situations, as well, and need more than a CAM system to help them out of their difficulties. However, when the system planner has actively begun consideration of competing systems and has his economic factors well in hand, he will find the vendors of some help in making the economic analysis. Most NC and CAM system vendors have prepared typical economic analyses that serve as useful patterns in comparing the performance of their systems to alternative systems and methods.

The question of actual costs for the CAM system is a function of the time that you are buying and the ambition of your system. More small- and medium-sized corporations are getting into the business of vending CAM systems to the general industrial and electronics markets and, in a nascent industry such as this, it is difficult to predict the future of hardware prices, except to say that they will probably be lower than at present. The concept of the commercial viability of computer-aided manufacturing systems as products has only recently been adopted by a few corporations and, as more get into the business, there will be further pressure for lower prices. The reduction in the costs of computer hardware and peripheral devices will also make it possible for these companies to further reduce their prices as they begin to recoup some of their initial product development costs.

The principal contributor to the cost of a CAM system is the engineering of the system and the computer hardware and software. Currently available CAM systems are usually offered by corporations that are principally manufacturers of electronics manufacturing tools, such as wire termination machines component insertion machines, and so on. Their line of computer-controlled products exists (for them) primarily to support the sales of their manufacturing tools. The computer hardware and, generally, the software are purchased by these corporations from outside sources for inclusion in their systems and, although it represents the largest single cost component, it does not represent a significant profit item for them. Their profits are derived principally from the machines that are a part of the system. For this reason, there is little motivation for the present vendors of CAM systems to offer fully integrated CAM systems, except to the extent that the integration includes each of the manufacturing tools in their line. Thus even the most integrated commercially available CAM systems offer very little in the way of management information functions. The most inclusive systems provide data on the current production rates of the machines that they control and occasionally offer modest software for the generation of part programs intended for operation on these systems. Therefore, if you are planning a fully integrated system, you will have to expect to handle most of the engineering and software work yourself, either directly or through your own subcontracts.

Since this is the state of affairs (and will most likely continue to be the state of affairs for quite some time), you will have to acquaint yourself with the costs of the original development of such systems and with the techniques of managing and procuring original software systems and computer hardware. Efficient control over these phases of the development of your CAM system is very important to your economic performance since, even in the direct computer control systems, the cost of the computer hardware can represent 30 percent or more of the cost of the hardware for the complete system. For simple direct control systems, a generally reliable rule of thumb is that the software will cost at least as much as the hardware. You should therefore figure that your basic control software (exclusive of major management information, inventory management, and production control systems) will cost as much as the computer hardware and that the additional software for the integration of the total system will add significantly to this figure. Typical stand-alone management systems for any one of these management functions presently cost from $50,000 to $100,000.

There are many factors that must be considered in estimating the cost of a CAM system and even the best-laid plans may leave something out. It is therefore important that your budgets have some surplus and that they be developed with the best advice that is available to you. You should factor in the opinions of your own in-house engineering staff, your corporate accounting people, and the potential vendors for all or any part of the system.

PROCUREMENT

If you are considering only a modest CAM system, such as those that are presently available commercially, then you will probably discover that your present corporate procurement techniques for the purchase of conventional capital equipment may be entirely satisfactory. In most cases, these consist of direct engineering contact with vendors after an initial screening and approval of budgetary estimates of the cost of the hardware, solicitation of price quotations, and technical evaluations by your in-house engineering and manufacturing management personnel. The system specifications provided by the vendors for their own products can generally be used as the basis for your own specifications, either as is or modified to reflect any customizing that you may require. As long as the customizing is not extensive, these techniques should normally produce satisfactory results.

For ambitious CAM systems, the development of total system specifications and the overall system design will be your responsibility. If your corporation is equipped with adequate in-house system design capability, you should use it; however, you must be absolutely sure that it is indeed adequate for the problem at hand. For example, an in-house systems engineering staff whose principal functions are business procedures and software will rarely be equipped to handle the problems of specifying an ambitious CAM system. Knowledge of machine control and real-time programming techniques is not often abundant in the typical organization of this type. Similarly, the manufacturing engineering staff may also find itself lacking in the expertise in computer matters necessary to provide an adequate job of designing and specifying an ambitious CAM system (although they might be perfectly capable of handling a modest system). Nevertheless, each of these groups may attempt to corner more than their fair share of control over the final design of the system. In general, when utilizing in-house personnel from nonmanufacturing organizations for these functions, it is more efficient to select the design team from among these groups yourself and to insist that they be responsible only to the system manager for the duration of the implementation plan.

If you are, for one reason or another, unable to establish an adequate design and planning staff from in-house personnel, you may find that consulting organizations or systems houses (companies whose principal work is the development of custom-made computer systems) can be of great assistance to you. If you have no in-house capabilities, this may be the only feasible route available to you. These organizations can be used in several ways. The most direct way is in the role of prime contractors for the system. One such organization could be employed as the prime contractor and still another organization as a contract monitor. The principal advantage to you is the availability of experienced advice and guidance without the need for your corporation to build up an extensive staff. If the nature of your system is such that continued assistance is required during the operation of the system, it may be obtained by keeping these corporations or consultants on a retainer or by actually hiring personnel from these organizations. This is not an uncommon practice, although it is not looked on favorably by the organizations that lose these personnel.

You may, as we have said, entertain bids from these oragnizations to serve as prime contractors or the actual developers of your system. Unless your system is at least moderately ambitious or represents some amount of follow-on work for the bidders, you must expect that they will keep their proposal work investment as small as they reasonably can. In these cases, their proposals will consist principally of brief descriptions of the system and their price quotation. If, of course, you already have an adequate specification and system design on which their quotations are based, this is not a serious problem. If this is not the case, it is important that you have a better description of exactly what you are buying. Therefore, after an initial screening of potential vendors, you should seriously consider the idea of partially funding the proposal work and insisting that the bidders submit thorough system designs and implementation plans along with their price quotations. You should make it clear that you will have rights to the designs that result from this work at least to the extent that you may employ them for your own use (except, of course, where prior patent and license agreements and similar legal restrictions prohibit this). In this way, you can freely select both your prime contractor and your technical design, independently of the relation between the technical design and the bidder who is finally selected. This approach will motivate bidders to provide you with better ideas, advance your design work, and give you a better picture of the capabilities of each bidder. The resulting quotations and schedules will also be more accurate, as they will be based on a more thorough investigation of the work to be done.

The placement of the final order should be for a specification that your corporation has developed, using the inputs from the bidders. This should incorporate all of the technical ideas developed during the initial bidding phases which you wish to have incorporated in your system. Thus there will be several phases to your procurement cycle:

- Initial screening of vendors
- Solicitation of detailed technical proposals (possibly funded)
- Preparation of a detailed technical specification and implementation plan
- Solicitation of price quotations based on the final specification
- Selection of the actual prime contractor and other vendors

Your technical specification should be made as comprehensive and detailed as it reasonably can. Aside from defining the performance characteristics of the system, it should also include a specification of the tests that will be performed to ensure that the performance characteristics are actually met by the completed system. These are generally called "acceptance tests."

The problem of CAM system procurement is a very complex one and it cannot be easily handled in a short period of time. Therefore, when you set out in your procurement, be sure to allocate enough time for all your potential vendors to prepare properly considered proposals and quotations and for your corporation to thoroughly evaluate them. Start early enough and be wary of external schedule constraints imposed on the development of your system. These generally come from situations in which the CAM system is a part of gearing up for a large manufacturing contract that your corporation expects. Often, these external schedule constraints are imposed with too little consideration of the problems of implementing the CAM system and they may present you with severe schedule problems of your own. "Haste makes waste" is particularly true in the context of designing and planning complex computer systems.

Chapter Eight

Trends

Predicting the future is a tricky business. Those predictions generally considered most accurate are the vague ones that no one understands, except in the context of hindsight or rueful experience. The predictions that everyone understands immediately are the sweeping ones which spell out the details of the future with great clarity and surety. They rarely come true. The present value of either kind of prediction is doubtful, and therefore we try to avoid them.

We discuss in this chapter some of the trends that are presently in evidence which have a bearing on the subject of computer aids to the electronic manufacturing process. It appears to be more useful to attempt to measure the present value of the "derivatives" of factors that have a bearing on this subject. Perhaps then we may find ways of taking advantage of certain trends, of accelerating those which seem to promise certain benefits, and of decelerating those which prejudice the goals that we may set for ourselves.

There are three major spheres to be considered: the economic sphere, the technological sphere, and the organizational sphere. Several interesting trends are taking place in each of these spheres which may, from present indications, have some future effect on electronics manufacturing and on the lives of social and industrial enterprises which now depend on electronics or which may come to depend on electronics in the future. Of the three spheres, the economic one is fraught with the greatest uncertainty. The technological sphere is the simplest to measure since its effects are based on the publicly expressed goals of people and companies in the field and are manifested by frequent announcements of new developments in the trade press. In the organizational sphere, things move rather more slowly and good statistical data are lacking, but trends are nevertheless apparent.

THE ECONOMIC SPHERE

In the economic sphere, three trends are readily apparent:

1. The costs of hardware are being reduced.
2. The costs of labor are increasing.
3. The capital markets have become more conservative about investing money in ventures based on computers.

The reduced costs of hardware bear on the computer-aided manufacturing of electronic gear in a number of ways. The most widely discussed factor is the greatly reduced cost of the computer, as typified by the minicomputer. With regard to CAM, this means that the cost of hardware for these systems will probably be less in the future than it is now. Not only are the costs of the computers going down, but the costs of electronic componentry for interfaces and peripheral equipment are also going down. Almost every manufacturer of minicomputers is in a state of imminent or recent announcement of reduced prices or of less expensive models of their products, in spite of general inflation in other areas of the economy.

The principal question here is whether such reductions can continue indefinitely. The prices of many minicomputers are now down to the level of a moderately expensive personal automobile. Without actually predicting it, there is the possibility that too great a reduction in the price of computers could lead to the kind of situation that many find in the automotive market, namely, less than satisfactory service facilities and reliability. If the users of computers can tolerate such a situation, prices may drop to incredibly low levels. However, there are indications that some users of computers are not satisfied with the present levels of service provided by manufacturers.

A principal factor in the reduced cost of the computer was the original equipment manufacturer (OEM) market. This market consisted of corporations that bought small computers in large quantities for inclusion in systems that were otherwise generally of their own manufacture. In return for lower prices, they gave up certain services from the manufacturer of the computer. The income from this source also helped the manufacturer to reduce his prices to the general single-system user. However, recent economic conditions have led to a retardation of the OEM markets and have motivated more manufacturers to concentrate to a greater extent on another segment of the market, the "end user." The end user frequently depends to a greater extent on the manufacturer for service and support in the use of the computer—services that the manufacturer may find difficult to supply within the structure of future lower prices. A great number of applications of the minicomputer in computer-aided manufacturing fall in the end-user area—the system custom designed for a particular manufacturing operation. Thus the CAM user may find himself in the position of paying very attractive prices for good hardware, but receiving very little in the way of needed services and having to provide them for himself. The total economic impact of these considerations is unclear. It depends on the relative effect of the reduced cost of the computer versus the cost to the user in providing his own services.

There is a way in which the trend to reduced prices may be turned to another advantage, although its implementation is largely in the hands of the computer manufacturer. This method is to divert a portion of the future reduction in the cost of components to providing more capable computers and improved services for the same total cost. Component developments tend to maintain or improve the reliability of computers, so that present levels of maintenance service should at least retain their present adequacy. The result should be equipment of improved capability, but at a price that should allow the manufacturer to provide better services to the end user. Several of the most prominent manufacturers of minicomputers are starting out in this direction by using reduced cost components to bring out virtual copies of their standard computers at reduced prices. These are very capable machines and are sold at prices that approximate the cost of the programmable controllers of a few years ago. This is half the right idea, and it is hoped that their next step will be to add new features to these machines or to improve their processing speeds, while holding the line on prices and improving services. The buyer of computers can accelerate this trend by asking the manufacturers for this kind of approach and by buying the more capable versions of their products.

The rising cost of labor is frequently treated as a motivation for the application of computers to manufacturing processes. However, in some short-term situations the cost of labor mitigates against the use of computers since a good deal of the initial cost of computerization is represented by engineering labor costs. On the whole, these engineering labor costs are rising as rapidly as general labor costs and are built on a higher base to begin with. If there are difficulties in raising capital for the installation of computer-aided manufacturing systems, it may be more fiscally viable for a corporation to continue to rely on present manufacturing methods rather than borrow or otherwise raise capital for the installation of a system. Although the eventual cost may be greater, the impact of debt financing or dilution of corporate stock may be deemed unacceptable. Many corporations considering the development of CAM systems for the first time have been somewhat taken aback by the engineering costs involved. If one subscribes to an inflationary psychology, one can only believe that the situation will get worse before it gets better. Present signs seem to indicate that quite some time will pass before there is a significant retardation in engineering labor costs.

One cannot expect to retard this trend by attempting to employ unusually cheap labor for the development of these systems. Their development, as the reader will no doubt appreciate from the technical considerations outlined in the previous chapters, is not a job for technical or managerial neophytes, and the utilization of low-grade engineering labor in the development of these systems may often lead to overall costs that are much greater than anticipated.

The possibility of eventual development of off-the-shelf systems is one way in which the effects of rising engineering labor costs can be attenuated. However, as we have pointed out once before, there are many technical problems standing in the way of this, notably, the unique needs of individual manufacturing organizations and the impact of these needs on the system designs. The users of CAM systems can aid this development by relaxing their requirements in ways that permit greater commonality between installations of commercially available CAM systems. For example, adoption of standard NC program formats, management information report formats, wiring lists, and part number structures would help greatly. Unfortunately, the prospects for such general cooperation are rather dim. There will always be situations in which special technical considerations are necessary in a particular manufacturing installation. It would be a start in the right direction, however, if this could be done wherever possible.

Another approach to attenuating costs would be to share the benefits of the system with the original manufacturer. This may be especially useful when dealing with custom-built systems. If the accounting problems can be surmounted and if the vendor can finance his end of the arrangement, a percentage of the cost savings attributable to the use of the system could be returned to the vendor over an initial period of its operation. If an equitable arrangement can be reached, the initial cost of the system can be reduced. If the system is amenable to general use as an eventual off-the-shelf product by the vendor, he can recover some of the development costs of his own product in the initial system. While this approach may prove very attractive, there are several problems. Chief among these is that of setting up an equitable accounting arrangement for the measurement and disbursement of the benefits gained from the operation of the system.

The recently acquired conservative attitude in the capital market regarding computer-based ventures represents a force retarding the general development and distribution of CAM systems. Although in the long run it will no doubt contribute to the general strength of the industry and the quality of corporations engaged in this business, in the immediate future it will make it more difficult for individual users to find viable vendors for the development of custom built CAM systems.

The failure of many computer-based ventures to live up to their original promise is the principal cause for this conservative attitude. There are many reasons for this failure. One is the tendency for companies engaged in computer-related activities, particularly software, to bury their development costs in capitalization. This tends to create a rosy financial picture on the books of the corporation, but it disguises the effect of rapidly changing technology on the worth of these capital assets and tends to put off discovery of the fact that many of the programs that these corporations are purveying are not worth the paper they were listed on—at least in the eyes of potential buyers. Capitalized software and computer designs do not always wear out in ways that are properly considered by standard accounting methods. It is often the case that they are obsoleted by the development of new technologies, new ways of doing operations, and new styles. Software and computers can literally go out of style even though they remain as capable as they ever were of performing their functions. Thus the returns on capitalized development costs tended not to materialize, and investors became generally disenchanted.

In the minicomputer market, although predictions for the total dollar value of the market were no doubt not too far off, too many companies tried to get into the business without realizing that their principal problem was simply to get enough exposure to potential customers. They failed to consider adequately the marketing aspect. To develop a minicomputer design and to set up an adequate production facility is not too difficult a task, given sufficient resources. Almost every one of the minicomputer designs has something useful to offer. But many of them failed to catch on because no one outside the inner circles of the computing community had ever heard of them. The largest users of computers tended to evaluate and buy only from the limited number of better known manufacturers. To the potential user of CAM systems, this trend might be interpreted as a warning to buy their computer equipment from relatively strong manufacturers and to avoid the products of less well-established firms. There are, of course, exceptions to this rule (we discussed some of them in Chapter 7) but in the majority of the cases, you will find it wise to pursue this policy.

This policy is quite difficult to follow in the case of general system engineering and software since there are no proper analogues. That is, there is no set of well-established software and systems engineering firms that can be viewed in the same light. The longevity of the vendor in this area is, however, not so critical as in the case of the computer manufacturer.

If the user gets proper documentation and if the vendor survives through the system burn-in period, then the user can consider himself relatively free of critical dependence on the original vendor. The key to this, however, is the quality of the documentation left behind and the degree to which the user understands the system that he purchased.

In my opinion, the conservatism of the capital market is a trend to be welcomed by the user of CAM systems. While it presents some difficulties at this time (and contributes to the capital-raising problems of the user as well as many of the vendors), the ultimate result should be a stronger and more capable group of corporations providing services in this area—one on which the users can depend with greater assurance of satisfactory performance.

THE TECHNOLOGICAL SPHERE

In the technological sphere there are two major trends that we consider:

1. The popular concept that the minicomputer can be treated as just another system element
2. New ways of viewing the minicomputer in various applications and of using the minicomputer in modern system configurations

At least in terms of the economics of hardware cost, the minicomputer can be treated as just another element in the system. Many consider this to be one of the major attractions of minicomputers. However, there is a very important consideration that makes this conception less than true and tends to mitigate against its eventual complete realization. This is the matter of programming and software development. These machines still require the talents of people capable in assembly language programming, especially for their use in CAM systems.

Some commentators in the computing profession advocate solutions to this problem through the development of computer languages that are easier to use and through the use of large-scale integration subsystems within the computer to handle major commonly used functions. I am personally skeptical of the ultimate success of these approaches. Languages that are easier for humans to use require enhanced computer resources for their translation into machine language, either in the computer system that hosts the final object program or in separate systems used only for the preparation of programs. Although a great deal of research is being devoted to the development of language translators that generate very efficient code, present indications are that it will be quite some time before automatic translators will be able to translate any procedure-oriented language into object code as efficient as that produced by a moderately talented computer programmer who pays proper attention to his work. When employing high-level languages, both the requirement for enhanced computer resources and the inefficiency of the code generated work against the inherent economic advantages of the minicomputer.

The use of LSI will certainly contribute to a reduction in cost for LSI equivalents of present computer architectures and will bring about certain advantages that are derived from reduced physical size and increased modularity, such as faster processing speeds and faster fault diagnosis and repair. But the idea that LSI computers will be significantly easier to program, in terms of the economic resources required for this task, is difficult to accept. We must consider that each application, in spite of efforts to standardize CAM practices and formats, will tend to have its peculiar problems which require special attention. It is here that programming is very important. Having an LSI element in a minicomputer that represents, for example, a preprogrammed input/output facility, will be useful if the processing problem at hand can be made to conform to the capability of this element. But if the problem cannot be made to conform in every detail, the user may find that he is left with a tricky problem.

An analogous situation is found in many cases in which the vendor of the computer has attempted to provide similar capabilities through the standard software that he supplies with his product. Consider, for example, the operating systems that many computer manufacturers offer with their hardware. To the best of my knowledge, no one has surveyed the extent to which these operating systems have been used and found to be of any value beyond providing the user with a few relatively straightforward services (in spite of the fact that some of them are designed to provide rather sophisticated general-purpose services). It has been my personal experience that in many cases the use of these operating systems has been rejected by the designers of custom-built systems. The principal reasons for this rejection have been either that the operating systems were so wasteful of computer resources (especially core memory) that they were not economic to use or that they lacked features required by the problem at hand which were difficult or impossible to incorporate into the existing structures of the operating system. In many other cases, the users of these systems found it necessary to make certain modifications to the systems. They were then faced with the problem of learning about the internal structure of the operating systems so that they could make these modifications. (I should except from these remarks some of the operating systems designed for computerized direct digital control of processes. These are generally more useful in their original forms, although the problems faced by the designers of these systems were much more standardized and amenable to parametric solutions than the problems faced in the development of a CAM system.) Inasmuch as LSI techniques are essentially envisioned as providing much the same features that the software operating systems and high-level languages were intended to provide, it is difficult to understand how they will resolve the economic problem of programming in a better fashion than the software systems, except to perform the same unsatisfactory services at greater speed. Furthermore, it is somewhat more difficult to modify an LSI chip than it is to modify a computer program.

Perhaps future technologies and improved classification and understanding of CAM processing problems will resolve this problem. At the moment, however, it seems that the best tactic for the user and potential user of CAM systems is to foster the development of talented minicomputer programmers, either through alliances with capable software firms or by developing these talents among the in-house engineering staff. These programmers must be knowledgeable in hardware, both electrical and mechanical, and should understand production control systems, numerical control techniques, and many of the basic principles of business data processing (especially inventory management). A knowledge of programming is becoming increasingly important to the professional stature of the engineer. A good deal of the current pressure in this direction is toward learning languages which aid in the tasks that are conventionally deemed proper for the practicing engineer, such as BASIC and FORTRAN, which are valuable in engineering calculations. More pressure should be directed toward learning the techniques of assembly language programming of small computers.

Several interesting ways of using the minicomputer have developed in recent years and hold promise for appreciable benefits in the future, especially in the area of computer-aided manufacturing systems. One of these is the use of modest minicomputer systems in stand-alone configurations, preprogrammed for particular applications that do not involve the direct control of equipment or machine tools. Examples of these are payroll systems, production control systems, and inventory systems. Such systems are engineered for use by personnel without formal training in computer programming or operation. Such a system can generally serve the needs of small- and medium-sized corporations and separate departments within larger corporations, and represent a much lower total investment than might be incurred through the acquisition of similar data processing capabilities by renting or leasing general-purpose business data processing systems.

Another interesting use of the minicomputer is as the principal component in the distributed computer system, which we discussed in Chapter 4. There is now a great deal of research interest in such system structures, and minicomputer manufacturers are taking note of this use of their products by offering optional peripheral gear that facilitates the interconnection of many minicomputers in such systems. In many applications, the principal technical problem with the distributed computer system is partitioning the work to be done by the computers and distributing the work throughout the heirarchy of computers in a way that provides the most economical utilization of their capabilities. For the most part, CAM systems for electronics manufacturing do not suffer from this problem since the lines for division of labor among the computers can be drawn rather easily. Each element of the manufacturing problem is represented by a single machine (component insertion, wire termination, etc.) and the control of any single class of machines (up to the limitations of the actual control minicomputer) can be assigned to a single computer. Inventory and production control can be separately handled by one or two other minicomputers. If development of manufacturing data bases is to be handled by a minicomputer system, it can be tied in with the network of control minicomputers in a straightforward manner.

The advantages that this approach brings to the field of CAM are modularity, flexibility, ease of system expansion, and reliability. With this kind of design it is much easier to start with a modest system and to expand its capabilities without having to invest in a large amount of spare computer resources in order to anticipate future needs. Potential users of CAM systems can accelerate this approach to system design by recommending to the bidders on their systems that they consider the feasibility of this approach, especially if these systems may someday be relatively ambitious in scale.

THE ORGANIZATIONAL SPHERE

Our concerns here are how corporations are organizing themselves in order to carry out their functions and the effect that automation is having on their organization. The principal question here is that of centralization versus decentralization—whether the corporation tends to be more closely controlled at its central headquarters or whether more control and decisions are handled farther out in the organizational structure, at such points as the manufacturing or the procurement operations.

In the early 1960s, many business analysts and computer people predicted that the use of the computer would reduce the need for middle levels of management. The thesis was that increased data handling speeds and the use of computers for the analysis of this data would make centralized and automatic decision making more feasible than was then the case. The effect of the computer on organizational structure is still a subject of some controversy, for there is now a relatively widespread opposite opinion that the effect of the computer will be to induce decentralization. Those who hold the centralization theory seem to favor the larger computer installations and consider the minicomputer only for local control and modest scientific applications. Those who favor the decentralization theory seem to have a better opinion of the capabilities of the minicomputer and believe that the natural tendency of a corporation is to decentralize, if it is technically feasible for it to do so. All of the evidence is yet to come for this question, but there seem to be indications that the decentralization theorists will win out in the long run. Among these indications is the absence of a great rush to centralize corporate structures. We discussed, in Chapter 6, some of the advantages of decentralization in which the minicomputer is used to provide local processing of information. The total question must consider three major factors: the gathering and transmittal of information, the actual processing of the information, and the making and implementation of decisions based on the information. Within the scope of present technology, local handling of all three factors is the least expensive and most responsive solution to the problem, at least in the case of tactical decisions. The original proponents of the centralization theory did not have the minicomputer to factor into the considerations that led to their views. With present technology, centralization seems to be enormously expensive in terms of hardware, system design, and development; its return over the local automation approach appears marginal in all but the most optimistic estimates.

In the final analysis, the people managing the corporation at all levels are the key to successful and profitable operation. Making their jobs easier is the proper application of computers. This does not, in my estimation, include removing from them the responsibility for the important tasks of decision-making and evaluation of their own operations. The most fruitful applications of computers are those in which the computers handle the quantitative data and present it to the managers in ways which aid them in making their decisions.

Bibliography

The following books are recommended as sources of technical information about the design, programming, use, and economics of computers:

Hellerman, Herbert, *Digital Computer System Principles,* McGraw-Hill, New York, 1967.

Maurer, Ward Douglas, *Programming: An Introduction to Computer Languages and Techniques,* Holden-Day, San Francisco, 1968.

Richards, R. K. *Electronic Digital Systems,* Wiley, New York, 1966.

F. Sharpe, William, *The Economics of Computers,* Columbia University Press, New York, 1969.

The following proceedings and magazines are recommended as continuing sources of information about the application of computers to electronic manufacturing:

Computer Design, 221 Baker Avenue, Concord, Mass. 01742.

Control Engineering, 466 Lexington Avenue, New York, N.Y. 10017.

Datamation, 35 Mason Street, Greenwich, Conn. 06830.

Electronic Packaging and Production, 222 West Adams Street, Chicago, Ill. 60606.

Proceedings of the National Electronic Packaging and Production Conferences, Industrial and Scientific Conference Management, Inc., 222 West Adams St., Chicago, Ill. 60606.

Proceedings of the Spring and Fall Joint Computer Conferences, AFIPS Press, 210 Summit Avenue, Montvale, N.J. 07645.

Index

Acceptance tests, 233
Actuators, selection of, 128-129
Addressing, 67
Architecture, advanced, 94-98
 classical, 64
Arithmetic, 42, 141
Artwork, 22
Assembler (program), 91, 100, 105

Background programs, 111, 168
Bill of materials, 147, 195
Binary coded decimal (BCD), 141
BTR systems, 133
Business data processing facilities, 187, 200
Byte, 67

Calculations, nature of, 42-43
CAM systems, acceptance, 218
 advantages, 205
 planning, 224
 procurement, 231
Card punches, 93
Card readers, 93
Centralization, 242
Central processing unit, 40-41, 64
COBOL, 107
Compilers, 91, 107
Component insertion machines, 34
Computers, business and scientific, 41
 classical structure, 40-43

design and packaging, 151
placement in the organization, 199
procurement, 47
selection, 123
Computer responsibility, 130
Computer system architecture, 63
Computer-aided design systems, 15-18
Computer requirements, 157-163
Computerized organization, conventional,
 197
Continuous System Modeling Program
 (CSMP), 19
Contouring, 27
Control console, 155
Controller, absolute, 30
 incremental, 25, 30
 programmed, 47
Contract monitors, 232
Cost, 228
 computer hardware, 231
 hardware, 235
 labor, 236
 minicomputer, 235
 principal contributors to, 230
Cost-effectiveness analysis, 202

Data flow, 192-197
Data collection, 186
 time lag in, 186
Data concentrators, 58

Dead band, 30, 131
Decentralization, 242
Decentralized computer structures, 76-77
Design objectives, 121
Digitizers, 23-24
Direct access device, 93
Direct memory access (DMA), 41, 86-88
Directories, 149
Disk memory, 92
Distributed computer system, 54-55, 241
Distribution systems, 11-12, 37-38
Documentation, 9, 13
Drafting machines, 21-23
Drills, printed circuit board, 34
Drum memory, 92
Dynamic range of control loop, 134, 140

Economics, direct computer control versus
    NC, 195
Effect of CAM system, 194, 196
Electronic Circuit Analysis Program (ECAP),
    17
Electronics manufacturing, characteristics
    of, 12-13, 116
    phases of, 6-12
Emulation, 75, 102
Engineering changes, distribution of, 117
Estimates, 160

Fabrication, 9-10
Fail soft, 55
File, access techniques, 149-151, 170
    chaining, 150
    cleanup, 150
Foreground programs, 111, 168
Formats of input commands and responses,
    156
FORTRAN, 100, 107

General Purpose System Simulator (GPSS),
    19
Graphic terminal devices, 59, 93, 155
Grouping of responsibility, 214

Human factors, 122, 154-157, 218
Hunting, 30

Illinois Institute of Technology Research
    Institute (IITRI), 25
Indexers, 135

location of, 152
Index registers, 43, 72
Information, as by-product of control, 189
    flow through organizational structure, 191
    generated by the CAM system, 189
    relation to management responsibilities,
        187
Information handling, centralized, 206
    decentralized, 209, 211
Input/Output Processors, 88-89
Input/Output, programmed, 78-81
Installation plan, 225
Instructions, 41, 64
    control, 70
    input/output, 71
    memory referencing, 69
    register referencing, 69
Instruction set, 65
Integrated electronics manufacturing sys-
    tem, 217
Interface control units, 134
    modularity of, 151
Interrupt, 43, 71, 81-86
Interrupt enable signal, 83
Interrupt priority signal, 83
Interrupt response program, 43, 85
Inventory file, 146, 195
    updating, 147
Inventory systems, 38

Keyboards, 156

Languages, assembly, 99, 102-106
    machine dependent, 99
    machine (object code), 99, 100-102
    procedure-oriented, 106-110
    procedure-oriented, advantages of, 109
Large scale integration systems, 239
Loading program, 105
Location counter, 70
Logic diagrams, 21

Machine conditions, 85
Macro, 105
Management, middle, 191, 197
Manager, effectiveness of, 215
    objectives and responsibilities of, 183
Manufacturing data base, 8-9, 145
    handling of, 144
Manufacturing operations, collection of

data, 188
coordination of, 183
evaluating performance of, 185
flow of information in, 185
Memories, associative, 96-97
buffer, 95
integrated circuit, 95
Memory, 40, 43
read-only (ROM), 74
Memory organizations, advanced, 94-97
Microprogramming, 71, 73-74
Minicomputer, 2, 6, 43-44
definition of, 46
effect on information handling, 214
stand-along configurations, 241
for business data processing, 187
Minicomputer systems, specially programmed, 201
Minimum computer responsibility control, 132
Modularity, 151
Motors, direct current, 28, 30
servo, 28, 30
stepping, 27-29, 126
stepping, hydraulic amplification of, 30

NC machines, absolute, 30
incremental, 25, 30
reasons for adopting, 33-34
NC systems, computerized, commercially
available, 133
Numerical control (NC), 10
applications, 34-35
Numerical control programs, standards for
representation, 145
Numerically controlled tools, 27-35

Object code, 105
Operating system, 62, 111, 240
Operator-computer communication, 155
Original equipment manufacturer (OEM),
235
Overlaying, 160, 170
Overstepping, 30

Packaging and general design, 151
Page, base, 68
fixed, 69
floating, 69
Paging of memory, 67-69

Part explosion file, 147
Part program, retention on disk memories,
154
transfer time of, 149
volume for wire termination, 148
volume occupied by, 148-149
Part program file, 177, 146
volatility, 146
Parts programming, 24-26
Peripheral devices, 41, 43, 89
Personnel, 227, 232
Planning, 220
Plotters, 23
Point-to-point, 27
Position error register size, 142-143
Positioning, absolute, 25
Positioning information, representation of,
159
Positioning rates, 129
Positioning systems, linear, 131
Position measurement, feedback loop of,
139
Price reductions, effects of, 47, 235
Prime contractors, 232
Printers, 93
Product design phase, 6-8
Productivity, 201
time lag, 204
Program counter (location counter), 70, 83
Program design, 163
Programs, 98
application, 60
assembly, 167
classes of, 167
editing, 111
executive, 167
input/output handlers, 167
interrupt response, 167
loading, 111, 168
overlay, 170
processing, 167
relocatable, 105
utility, 170
Program units, 105

Quickpoint language, 25

Random access device, 93
Rate control, 139
Real time, 49

Recorders, cartridge and cassette, 92
Register structure, 72
Registers, index, 43, 72
Reliability of computer, 120
Retrofitting, 133, 143
Return on investment, 182
Routine, 164

Sampled data system, 124-126
Servomotor, 28, 30
Simulation, 8, 18-19
Source code, 105
Sources of information, 221
Specifications, 233
Speed of computer, 119
Start/stop rates, 137-138
Stock systems, 11-12, 37-38
Stratification, horizontal, 191
    vertical, 191
Structure of management, 190
Subcontractors, monitoring and control of,
    228
Subroutine, 164
    functionally oriented, 165
Swapping, 62
Symbol table, 104
Systems, custom designed, 114
    multi- and parallel-processor, 97-98
    operating, 62, 111, 240
    process control, 51-55
    procurement of, 114-115, 231
    retrofit, 114
    software, 110-112
    stand-alone, 49-51

time-sharing, 17, 57-63
tracking, 55-57

Tape, conventional magnetic, 91
    incremental magnetic, 92
    punched paper, 91
Tape codes, standards, 32
Tape formats, 32
Tape readers, 154
Tape punches, 154
Teletypewriter (ASR33), 59, 90, 152
Testing, 10-11, 35-37
Time-sharing, economic aspects, 62
    services, 62
    systems, 17, 57-63
Trade conferences, 222
Trade magazines, 221
Training, in computer programming, 156
    management information, 157
Translators, 135
Trapping, 85

Unit record equipment, 21

Vendors, as sources of information, 224

Wire termination machines, automatic, 25-
    26, 34
    semiautomatic, 26, 35
Wiring, 25-26
Wiring lists, 8, 20
Word length, 41
Word size, instruction, 64
    memory, 64

DATE DUE

GAYLORD   PRINTED IN U.S.A